AIR QUALITY

Monitoring, Measuring, and
Modeling Environmental Hazards

AIR QUALITY

Monitoring, Measuring, and Modeling Environmental Hazards

Edited by
Marco Ragazzi, PhD

Apple Academic Press Inc. | Apple Academic Press Inc.
3333 Mistwell Crescent | 9 Spinnaker Way
Oakville, ON L6L 0A2 | Waretown, NJ 08758
Canada | USA

©2017 by Apple Academic Press, Inc.

First issued in paperback 2021

Exclusive worldwide distribution by CRC Press, a member of Taylor & Francis Group

No claim to original U.S. Government works

ISBN 13: 978-1-77-463038-9 (pbk)
ISBN 13: 978-1-77-188427-3 (hbk)

Library and Archives Canada Cataloguing in Publication

Air quality (Apple Academic Press)
Air quality : monitoring, measuring, and modeling environmental hazards / edited by Marco Ragazzi, PhD.

Includes bibliographical references and index.
Issued in print and electronic formats.
ISBN 978-1-77188-427-3 (hardcover).--ISBN 978-1-77188-428-0 (pdf)
1. Air--Pollution--Measurement. 2. Air--Pollution--Environmental aspects. I. Ragazzi, Marco, author, editor II. Title.

| TD890.A47 2016 | 628.5'30287 | C2016-901481-9 | C2016-901482-7 |

Library of Congress Cataloging-in-Publication Data

Names: Ragazzi, Marco, editor.
Title: Air quality : monitoring, measuring, and modeling environmental hazards / editor, Marco Ragazzi, PhD.
Description: Toronto : Apple Academic Press, 2016. | Includes bibliographical references and index.
Identifiers: LCCN 2016009305 | ISBN 9781771884273 (hardcover : alk. paper)
Subjects: LCSH: Air--Pollution--Measurement. | Air--Pollution--Environmental aspects.
Classification: LCC TD890 .A375 2016 | DDC 363.739/263--dc23
LC record available at http://lccn.loc.gov/2016009305

About the Editor

MARCO RAGAZZI, PhD

Marco Ragazzi has a PhD in sanitary engineering from Milan Polytechnic, Italy. The author or co-author of more than 500 publications (111 in the Scopus database), he is currently a member of the Department of Civil, Environmental, and Mechanical Engineering at the University of Trento, Italy. His research interests include solid waste and wastewater management, environmental engineering, and environmental impact risk assessment.

Contents

Part V: Agriculture

Acknowledgment and How to Cite

The editor and publisher thank each of the authors who contributed to this book. The chapters in this book were previously published in various places in various formats. To cite the work contained in this book and to view the individual permissions, please refer to the citation at the beginning of each chapter. Each chapter was read individually and carefully selected by the editor; the result is a book that provides a nuanced look at modeling air quality. The chapters included are broken into five sections, which describe the following topics:

- Chapter 1 analyzes some important perspectives of low-cost and high-density monitoring network for a more direct control of the human health risk from atmospheric macro-pollutants.
- Chapter 2 investigates what methods and approaches are commonly used in the published literature to characterize exposure levels from waste incinerators; it also assesses, through a computer simulation study, how the classification of the expected exposure level may change as a function of the method used to estimate it.
- Chapter 3 uses a sampling system coupled directly to aerosol spectrometers for the determination of fine and ultra-fine particles in the emissions of a waste incineration plant, suggesting that it is not a significant source of polychlorinated dibenzo-p-dioxin and polychlorinated dibenzofuran (PCDD/F) emissions or fine and ultra-fine particles.
- In chapter 4, the authors provide a detailed comparison of atmospheric dispersion modeling and a distance-based method to assess exposure to particulates from two municipal solid waste incinerators and explore issues of exposure misclassification.
- Chapter 5 provides a comparative analysis of the currently available technologies for measuring particulate releases to the atmosphere.
- Chapter 6 points out the role of PCDD/Fs monitoring in a case study related to a steel-making plant in a valley in the north of Italy, with the role of unconventional monitoring through characterization discussed in detail.
- Chapter 7 is an Italian case study whose PCDD/F deposition measurements are in progress using two types of deposimeters placed in a selected site, which discusses the seasonality and operativity of the plant.

- Chapter 8 provides initial estimates of regional air emissions generated by Pennsylvania-based shale gas extraction activities and associated ranges of potential regional monetized damages, which must be considered in the context of other external costs and benefits of shale gas extraction and use.
- Chapter 9 presents a methodology to study the role of urban street canyons in the stagnation of pollutants and to detect critical situations of exposure to air pollutants in a densely built area.
- Chapter 10 studies the causes responsible for the variability of levels of aerosol number concentration, black carbon, particulate matter, and gaseous pollutants at a selection of air-quality monitoring sites representative of different climate zones and urban environments in Europe.
- Chapter 11 investigates the Near-road Exposures and Effects of Urban Air Pollutants Study (NEXUS), including the respiratory health impacts of exposure to traffic-related air pollutants for children with asthma living near major roads in Detroit, Michigan.
- Chapter 12 presents extensive data on the characterization of two major crop-residue (paddy- and wheat-residue) burning emissio

List of Contributors

Shmuel Abramzon
RAND Corporation, 1776 Main Street, Santa Monica, CA 90407, USA

Giulio Angelucci
Waste Management Office, Autonomous Province of Bozen, Bozen, Italy

A. Alastuey
Institute for Environmental Assessment and Water Research (IDÆA-CSIC), Barcelona, Spain

Gianluca Antonacci
CISMA Srl, via Siemens 19, 39100 Bolzano, Italy

Saravanan Arunachalam
Institute for the Environment, University of North Carolina at Chapel Hill, 100 Europa Drive, Chapel Hill, NC 27517, USA

Danielle C. Ashworth
Department of Epidemiology and Biostatistics, MRC-PHE Centre for Environment and Health, Faculty of Medicine, Imperial College London, St Mary's Campus, Norfolk Place, London W2 1PG, UK

Stuart Batterman
Department of Environmental Health Sciences, School of Public Health, University of Michigan, Room 6075 SPH2, 1420 Washington Heights, Ann Arbor, MI 48109-2029 USA

Sarah Bereznicki
National Exposure Research Laboratory, United States Environmental Protection Agency, 109 T.W. Alexander Drive, Research Triangle Park, NC 27711 USA

Marco Brini
Minteos srl , Torino, Italy

Janet Burke
National Exposure Research Laboratory, United States Environmental Protection Agency, 109 T.W. Alexander Drive, Research Triangle Park, NC 27711 USA

Nicholas Burger
RAND Corporation, 1200 South Hayes Street, Arlington, VA 22202, USA

Beatrice Castellani
Consorzio IPASS Scarl, Via G. Guerra 23, 06127, Perugia, Italy

Mauro Chelodi
Dolphin srl , Trento, Italy

Alessandro Chistè
Civil Environmental and Mechanical Engineering Department, University of Trento, Trento, Italy

Marco Ciolli
Engineering Faculty, Civil and Environmental Department, University of Trento, Italy

Michele Cordioli
Department of Bio-Sciences, University of Parma, Parco Area delle Scienze 11/a, 43124 Parma, Italy and Regional Reference Centre Environment & Health, Regional Agency for Environmental Protection in Emilia-Romagna, Via Begarelli 13, 41121 Modena, Italy

Franco Cotana
CIRIAF, University of Perugia, Via G. Duranti 67, 06125, Perugia, Italy

Aimee Curtright
RAND Corporation, 4570 Fifth Avenue, Pittsburgh, PA 15213, USA

M. Dall'Osto
Institute for Environmental Assessment and Water Research (IDÆA-CSIC), Barcelona, Spain

Kees de Hoogh
Small Area Health Statistics Unit, Department of Epidemiology and Biostatistics, MRC-PHE Centre for Environment and Health, Faculty of Medicine, Imperial College London, St Mary's Campus, Norfolk Place, London W2 1PG, UK

A. M. Sánchez de la Campa
University of Huelva, Associate Unit CSIC "Atmospheric Pollution", Huelva, Spain,

J. de la Rosa
University of Huelva, Associate Unit CSIC "Atmospheric Pollution", Huelva, Spain

Giulio A. De Leo
Hopkins Marine Station and Woods Institute for the Environment, Stanford University, 120 Oceanview Boulevard, Pacific Grove, CA 93950, USA

Kathie Dionisio
National Exposure Research Laboratory, United States Environmental Protection Agency, 109 T.W. Alexander Drive, Research Triangle Park, NC 27711 USA

Paul Elliott
Small Area Health Statistics Unit, Department of Epidemiology and Biostatistics, MRC-PHE Centre for Environment and Health, Faculty of Medicine, Imperial College London, St Mary's Campus, Norfolk Place, London W2 1PG, UK

R. Fernández-Camacho
University of Huelva, Associate Unit CSIC "Atmospheric Pollution", Huelva, Spain

Mirko Filipponi
CIRIAF, University of Perugia, Via G. Duranti 67, 06125, Perugia, Italy

Anna Font
MRC-PHE Centre for Environment and Health, King's College London, Strand, London WC2R 2LS, UK

Gary W. Fuller
MRC-PHE Centre for Environment and Health, King's College London, Strand, London WC2R 2LS, UK

Val Garcia
National Exposure Research Laboratory, United States Environmental Protection Agency, 109 T.W. Alexander Drive, Research Triangle Park, NC 27711 USA

Eleonora Girelli
Civil Environmental and Mechanical Engineering Department, University of Trento, Trento, Italy

Y. Gonzaléz
Izana Atmospheric Research Centre, AEMET, Associate Unit CSIC "Studies on Atmospheric Pollution", Santa Cruz de Tenerife, Canary Islands, Spain

Anna L. Hansell
Small Area Health Statistics Unit, Department of Epidemiology and Biostatistics, MRC-PHE Centre for Environment and Health, Faculty of Medicine, Imperial College London, St Mary's Campus, Norfolk Place, London W2 1PG, UK

R. M. Harrison
School of Geography, Earth and Environmental Sciences, University of Birmingham, Birmingham, UK

David Heist
National Exposure Research Laboratory, United States Environmental Protection Agency, 109 T.W. Alexander Drive, Research Triangle Park, NC 27711 USA

C. Hueglin
Laboratory for Air Pollution and Environmental Technology, Swiss Federal Laboratories for Materials Science and Technology (EMPA), Dubendorf, Switzerland

Gabriela Ionescu
Power. Energy and Use Department, Politehnica University of Trento/ research fellow at Civil Environmental and Mechanical Engineering Department, University of Trento, Trento, Italy

Vlad Isakov
National Exposure Research Laboratory, United States Environmental Protection Agency, 109 T.W. Alexander Drive, Research Triangle Park, NC 27711 USA

Paolo Lauriola
Regional Reference Centre Environment & Health, Regional Agency for Environmental Protection in Emilia-Romagna, Via Begarelli 13, 41121 Modena, Italy

Aviva Litovitz
RAND Corporation, 1776 Main Street, Santa Monica, CA 90407, USA

Marianna Marconi
Civil Environmental and Mechanical Engineering Department, University of Trento, Trento, Italy

Luca Marmo
Science of Materials and Chemical Engineering, Technical University of Torino, Italy

T. Moreno
Institute for Environmental Assessment and Water Research (IDÆA-CSIC), Barcelona, Spain

Elena Morini
CIRIAF, University of Perugia, Via G. Duranti 67, 06125, Perugia, Italy

Andrea Nicolini
CIRIAF, University of Perugia, Via G. Duranti 67, 06125, Perugia, Italy

Massimo Palombo
CIRIAF, University of Perugia, Via G. Duranti 67, 06125, Perugia, Italy

Steve Perry
National Exposure Research Laboratory, United States Environmental Protection Agency, 109 T.W. Alexander Drive, Research Triangle Park, NC 27711 USA

J. Pey
Institute for Environmental Assessment and Water Research (IDÆA-CSIC), Barcelona, Spain

A. S. H. Prévôt
Laboratory of Atmospheric Chemistry, Paul Scherrer Institut, 5232 Villigen PSI, Switzerland

X. Querol
Institute for Environmental Assessment and Water Research (IDÆA-CSIC), Barcelona, Spain

P. Quincey
Analytical Science Team, National Physical Laboratory, Hampton Road, Teddington, Middlesex TW11 0LW, UK

Elena Cristina Rada
Civil Environmental and Mechanical Engineering Department, University of Trento, Trento, Italy

Marco Ragazzi
Civil Environmental and Mechanical Engineering Department, University of Trento, Trento, Italy

Andrea Ranzi
Regional Reference Centre Environment & Health, Regional Agency for Environmental Protection in Emilia-Romagna, Via Begarelli 13, 41121 Modena, Italy

Prashant Rajput
Physical Research Laboratory, Ahmedabad 380 009, India

C. Reche
Institute for Environmental Assessment and Water Research (IDÆA-CSIC), Barcelona, Spain and Institut de Ciencia i Tecnologia Ambientals (ICTA), Universidad Autònoma de Barcelona, Barcelona, Spain

S. Rodríguez
Izana Atmospheric Research Centre, AEMET, Associate Unit CSIC "Studies on Atmospheric Pollution", Santa Cruz de Tenerife, Canary Islands, Spain

Federico Rossi
CIRIAF, University of Perugia, Via G. Duranti 67, 06125, Perugia, Italy

Constantine Samaras
RAND Corporation, 4570 Fifth Avenue, Pittsburgh, PA 15213, USA

Manmohan Sarin
Physical Research Laboratory, Ahmedabad 380 009, India

Marco Schiavon
Fondazione Trentina per la Ricerca sui Tumori c/o University of Trento, Department of Civil, Environmental and Mechanical Engineering, via Mesiano 77, 38123 Trento, Italy

Deepti Sharma
Punjabi University, Patiala 147 002, India

Darshan Singh
Punjabi University, Patiala 147 002, India

Michelle Snyder
Institute for the Environment, University of North Carolina at Chapel Hill, 100 Europa Drive, Chapel Hill, NC 27517, USA

Maurizio Tava
Environmental Protection Agency, Trento, Italy

Werner Tirler
Eco Research Srl, Bolzano

Mireille B. Toledano
Department of Epidemiology and Biostatistics, MRC-PHE Centre for Environment and Health, Faculty of Medicine, Imperial College London, St Mary's Campus, Norfolk Place, London W2 1PG, UK

Marco Tubino
Civil Environmental and Mechanical Engineering Department, University of Trento, Trento, Italy

Alan Vette
National Exposure Research Laboratory, United States Environmental Protection Agency, 109 T.W. Alexander Drive, Research Triangle Park, NC 27711 USA

M. Viana
Institute for Environmental Assessment and Water Research (IDÆA-CSIC), Barcelona, Spain

Pietro Zambelli
Engineering Faculty Civil and Environmental Department, University of Trento, Italy

Dino Zardi
University of Trento, Department of Civil, Environmental and Mechanical Engineering, via Mesiano 77, 38123 Trento, Italy

Introduction

Air quality is at the forefront of the world's attention. The international community is seeking to expand global regulation for air-quality management, which must be based on accurate understanding of the health and overall environmental implications of macro- and micro-pollutants. Air pollution is a major environmental risk to health. By reducing air pollution levels, countries can reduce the burden of disease from stroke, heart disease, lung cancer, and both chronic and acute respiratory diseases, including asthma. The lower the levels of air pollution, the better the cardiovascular and respiratory health of the population will be, both long- and short-term.

Policies and investments supporting cleaner air quality focus on reducing emissions from various sources, such as waste incineration, industrial emissions, urban air pollution, and air pollution from agricultural sources. Reducing air pollution also reduces emissions of carbon dioxide and short-lived climate-change pollutants, such as black carbon particles and methane, thus contributing to the near- and long-term mitigation of climate change.

Ultimately, however, protecting and improving air-quality requires knowledge about the types and levels of pollutants being emitted. It also requires the best possible measurement and monitoring capabilities. The articles in this volume have been chosen as a foundation for monitoring, measuring, and modeling air pollution.

—Marco Ragazzi

Low cost sensors open to a new vision of the air quality control. Their performances allow for a new strategy closer to the population and its health. Critical situations that cannot be seen with conventional approaches can be managed quickly through an original network of sensors. In Chapter 1, Rada and colleagues present the preliminary steps of an integrated sensor based research. In particular, criteria for sensor selection referred to significant case-studies are discussed.

Incineration is a common technology for waste disposal, and there is public concern for the health impact deriving from incinerators. Poor exposure assessment has been claimed as one of the main causes of inconsistency in the epidemiological literature. In Chapter 2, Cordioli and colleagues reviewed 41 studies on incinerators published between 1984 and January 2013 and classified them on the basis of exposure assessment approach. Moreover, the authors performed a simulation study to explore how the different exposure metrics may influence the exposure levels used in epidemiological studies. Nineteen studies used linear distance as a measure of exposure to incinerators, 11 studies atmospheric dispersion models, and the remaining 11 studies a qualitative variable such as presence/absence of the source. All reviewed studies utilized residence as a proxy for population exposure, although residence location was evaluated with different precision (e.g., municipality, census block, or exact address). Only one study reconstructed temporal variability in exposure. Our simulation study showed a notable degree of exposure misclassification caused by the use of distance compared to dispersion modelling. The article suggests that future studies (i) make full use of pollution dispersion models; (ii) localize population on a fine-scale; and (iii) explicitly account for the presence of potential environmental and socioeconomic confounding.

Chapter 3, by Ragazzi and colleagues, presents the case study of a waste incinerator located in a region rich in natural and environmental resources, and close to the city of Bozen, where there are about 100,000 inhabitants. Local authorities paid special attention to the effect of the plant on human health and the surrounding environment. Indeed, among the measures adopted to control the emissions, in 2003 an automatic sampling system was installed specifically to monitor polychlorinated dibenzo-p-dioxin and polychlorinated dibenzofuran (PCDD/F) emissions during the complete operation time of the plant. The continuous sampling system was coupled directly to aerosol spectrometers for the determination of fine and ultra-fine particles in the emissions of the plant. The measurement results suggest that the waste incineration plant of Bozen is not a significant source of PCDD/F, or fine and ultra-fine particles. Immission measure-ments from other monitoring systems confirmed these results.

Research to date on health effects associated with incineration has found limited evidence of health risks, but many previous studies have been constrained by poor exposure assessment. Chapter 4, by Ashworth and colleagues, provides a comparative assessment of atmospheric dis-persion modelling and distance from source (a commonly used proxy for exposure) as exposure assessment methods for pollutants released from incinerators. Distance from source and the atmospheric dispersion model ADMS-Urban were used to characterise ambient exposures to particulates from two municipal solid waste incinerators (MSWIs) in the UK. Addi-tionally an exploration of the sensitivity of the dispersion model simu-lations to input parameters was performed. The model output indicated extremely low ground level concentrations of PM_{10}, with maximum con-centrations of $<0.01\,\mu g/m^3$. Proximity and modelled PM_{10} concentrations for both MSWIs at postcode level were highly correlated when using con-tinuous measures (Spearman correlation coefficients ~ 0.7) but showed poor agreement for categorical measures (deciles or quintiles, Cohen's kappa coefficients ≤ 0.5). To provide the most appropriate estimate of ambient exposure from MSWIs, it is essential that incinerator character-istics, magnitude of emissions, and surrounding meteorological and topo-graphical conditions are considered. Reducing exposure misclassification is particularly important in environmental epidemiology to aid detection of low-level risks.

The installation and operation of continuous particulate emission monitors in industrial processes has become well developed and common practice in industrial stacks and ducts over the past 30 years, reflecting regulatory monitoring requirements. Continuous emissions monitoring equipment is installed not only for regulatory compliance, but also for the monitoring of plant performance, calculation of emissions inventories and compilation of environmental impact assessments. Particulate matter (PM) entrained in flue gases is produced by the combustion of fuels or wastes. The size and quantity of particles released depends on the type of fuel and the design of the plant. Chapter 5, by Castellani and colleagues, provides an overview of the main industrial emission sources, a descrip-tion of the main types of monitoring systems offered by manufacturers and a comparative analysis of the currently available technologies for measuring dust releases to atmosphere.

PCDD/F emissions from steel making plants are characterised by conveyed and diffused streams. As a consequence, the conventional approach based on the control of emissions at the stack could not be sufficient to guarantee an adequate environmental protection. Chapter 6, by Rada and colleagues, focuses on the results of a recent multi-disciplinary study on a steel making plant located in the North of Italy in order to point out some control criteria based on unconventional monitoring. In particular, the role of deposimeters and soil characterisation is discussed in details.

As a case-study the local impact measurement of a sintering plant concerning the human exposure to air PCDD/F pollution is in progress, through the use of two types of deposimeters (conventional and wet & dry). Chapter 7, by Ragazzi and colleagues, deals with some preliminary results and some differences that must be taken into account when the two types of instruments are used..

In Chapter 8, Litovitz and colleagues provide a first-order estimate of conventional air pollutant emissions, and the monetary value of the associated environmental and health damages, from the extraction of unconventional shale gas in Pennsylvania. Region-wide estimated damages ranged from $7.2 to $32 million dollars for 2011. The emissions from Pennsylvania shale gas extraction represented only a few per cent of total statewide emissions, and the resulting statewide damages were less than those estimated for each of the state's largest coal-based power plants. On the other hand, in counties where activities are concentrated, NOx emissions from all shale gas activities were 20–40 times higher than allowable for a single minor source, despite the fact that individual new gas industry facilities generally fall below the major source threshold for NOx. Most emissions are related to ongoing activities, i.e., gas production and compression, which can be expected to persist beyond initial development and which are largely unrelated to the unconventional nature of the resource. Regulatory agencies and the shale gas industry, in developing regulations and best practices, should consider air emissions from these long-term activities, especially if development occurs in more populated areas of the state where per-ton emissions damages are significantly higher.

In Chapter 9, Schiavon and colleagues apply a modelling approach to an urban area, in order to study the effects of urban canopy in favoring critical situations of exposure to traffic induced air pollutants. The atmospheric dispersion of NOx, emitted by road traffic, was simulated inside the urban canopy layer by means of the COPERT algorithm and the AUSTAL2000 dispersion model. As expected, high concentrations occurred inside street canyons with consequences on the human exposure. The positive effect of traffic management options, such as incentivizing the public transportation and excluding the most pollutant vehicles from the circulation, was also investigated.

In many large cities of Europe standard air quality limit values of particulate matter (PM) are exceeded. Emissions from road traffic and biomass burning are frequently reported to be the major causes. As a consequence of these exceedances a large number of air quality plans, most of them focusing on traffic emissions reductions, have been implemented in the last decade. In spite of this implementation, a number of cities did not record a decrease of PM levels. Thus, is the efficiency of air quality plans overestimated? Do the road traffic emissions contribute less than expected to ambient air PM levels in urban areas? Or do we need a more specific metric to evaluate the impact of the above emissions on the levels of urban aerosols? Chapter 10, by Reche and colleagues, shows the results of the interpretation of the 2009 variability of levels of PM, Black Carbon (BC), aerosol number concentration (N) and a number of gaseous pollutants in seven selected urban areas covering road traffic, urban background, urban-industrial, and urban-shipping environments from southern, central and northern Europe. The results showed that variations of PM and N levels do not always reflect the variation of the impact of road traffic emissions on urban aerosols. However, BC levels vary proportionally with those of traffic related gaseous pollutants, such as CO, NO_2 and NO. Due to this high correlation, one may suppose that monitoring the levels of these gaseous pollutants would be enough to extrapolate exposure to traffic-derived BC levels. However, the BC/CO, BC/NO_2 and BC/NO ratios vary widely among the cities studied, as a function of distance to traffic emissions, vehicle fleet composition and the influence of other emission sources such as biomass burning. Thus, levels of BC should be measured at air quality monitoring sites. During morning traffic rush hours, a narrow variation in the N/BC ratio was evidenced, but a wide variation of this ratio was determined for the noon period. Although in central and northern Europe N and BC levels tend to vary simultaneously, not only during the traffic rush hours but also during the whole day, in urban background stations in southern Europe maximum N levels coinciding with minimum BC levels are recorded at midday in all seasons. These N maxima recorded in southern European urban background environments are attributed to midday nucleation episodes occurring when gaseous pollutants are diluted and maximum insolation and O_3 levels occur. The occurrence of SO_2 peaks may also contribute to the occurrence of midday nucleation bursts in specific industrial or shipping-influenced areas, although for PM, Black Carbon and particle number concentration at several central European sites similar levels of SO_2 are recorded without yielding nucleation episodes. Accordingly, it is clearly evidenced that N variability in different European urban environments is not equally influenced by the same emission sources and atmospheric processes. The authors conclude that N variability does not always reflect the impact of road traffic on air quality, whereas BC is a more consistent tracer of such an influence.

A major challenge in traffic-related air pollution exposure studies is the lack of information regarding pollutant exposure characterization. Air quality modeling can provide spatially and temporally varying exposure estimates for examining relationships between traffic-related air pollutants and adverse health outcomes. In Chapter 11, Isakov and colleagues used a hybrid air quality modeling approach to estimate exposure to traffic-related air pollutants in support of the Near-Road Exposures and Effects of Urban Air Pollutants Study (NEXUS) conducted in Detroit (Michigan, USA). Model-based exposure metrics, associated with local variations of emissions and meteorology, were estimated using a combination of the American Meteorological Society/Environmental Protection Agency Regulatory Model (AERMOD) and Research LINE-source dispersion model for near-surface releases (RLINE) dispersion models, local emission source information from the National Emissions Inventory, detailed road network locations and traffic activity, and meteorological data from the Detroit City Airport. The regional background contribution was estimated using a combination of the Community Multi-scale Air Quality (CMAQ) and the Space-Time Ordinary Kriging (STOK) models. To capture the near-road pollutant gradients, refined "mini-grids" of model receptors were placed around participant homes. Exposure metrics for CO, NOx, $PM_{2.5}$ and its components (elemental and organic carbon) were predicted at each home location for multiple time periods including daily and rush hours. The exposure metrics were evaluated for their ability to characterize the spatial and temporal variations of multiple ambient air pollutants compared to measurements across the study area.

In Chapter 12, Rajput and colleagues studied the characteristics and emission budget of carbonaceous species from two distinct post-harvest agricultural-waste (paddy- and wheat-residue) burning emissions from a source region (Patiala: 30.2°N, 76.3°E; 250 m amsl) in the Indo-Gangetic Plain (IGP), Northern India. The PM2.5 mass concentration varies from 60 to 390 $\mu g\ m^{-3}$ during paddy-residue burning (October–November) with dominant contribution from organic carbon (OC≈33%), whereas contribution from elemental carbon (EC) centres at ~4%. Water-soluble organic carbon (WSOC) accounts for about 50% of OC. In contrast, mass concentration of $PM_{2.5}$ during the period of wheat-residue burning (April–May) is significantly lower, varies from 18 to 123 $\mu g\ m^{-3}$ and mass fractions of EC and OC are 7 and 26%, respectively. The diagnostic ratios of OC/EC (11±2), WSOC/OC (0.52±0.02), nss-K+/OC (0.06±0.00) and ΣPAHs/EC (4.3±0.7 mg/g) from paddy-residue burning emissions are significantly different than those from wheat-residue burning (OC/EC: 3.0±0.4; WSOC/OC: 0.60±0.03; nss-K+/OC: 0.14±0.01 and ΣPAHs/EC: 1.3±0.2 mg/g). The emission budget of OC, EC and ΣPAHs from post-harvest agricultural-waste burning in the IGP are estimated to be 505±68 Gg/y, 59±2 Gg/y and 182±32 Mg/y, respectively. From a global perspective, crop-residue burning in Northern India contributes nearly 20% of both OC and EC to the total emission budget from the agricultural-waste burning.

PART I

OVERVIEW

CHAPTER 1

Perspectives of Low-Cost Sensors Adoption for Air Quality Monitoring

ELENA CRISTINA RADA, MARCO RAGAZZI, MARCO BRINI,
LUCA MARMO, PIETRO ZAMBELLI, MAURO CHELODI,
AND MARCO CIOLLI

1.1 INTRODUCTION

The environment has always been on the first step in the attention of advanced countries from many decades. In recent years the European Union regulation for air quality management has reached important results in term of exposure and health implications of macro and micro-pollutants and in terms of protection of the environment. However, many actions remain to be developed mainly in urban areas, but the general trend is towards an average improvement of air quality with positive consequence on the health of the population.

Perspectives of Low-Cost Sensors Adoption for Air Quality Monitoring. © Rada EC, Ragazzi M, Brini M, Marmo L, Zambelli P, Chelodi M, and Ciolli M. UPB Scientific Bulletin, Series D: Mechanical Engineering Journal **74,**2 (2012). *Reprinted with permission from the authors.*

Generally, the adopted regulations for air quality management are based on the concept of protecting the environment without facing with micro-scale critical situations, where human exposure to atmospheric pollutants can be inacceptable.

The monitoring and environmental warning systems today allow having some environmental information that is not sufficient or adequate for planning detailed corrective actions, or to quickly highlight critical situations potentially harmful to public health.

Air quality monitoring is a complex problem that requires the integration of multiple environmental information. Usually these data are coming from different environmental networks and often they are managed by different institutions. Therefore, sensor network, GIS models that indentify critical locations and Sensor Observation Service (SOS) that collect data and meta-data are used.

In the literature, three methods for interpolating air pollution data are available. The results from these can be combined with information from bioindicators [1-2].

- The Kriging interpolation method [3-4];
- Land Use Regression (LUR) method [4-5];
- Diffusion/ dispersion modeling of pollutants [6-7] and GIS technology [8-9].

In this frame, the Wireless Sensor Networks are effective means for monitoring in details environment pollution and life hazards, for example air quality in the cities and around storage and processing facilities such as ports, plants and dumps, fire warning through specific combustion gas detection, water pollution, dangerous or lethal gas warning in mining and oil industry [10-11].

Anyway, protocols and sensors are extremely new, and much research remains to be done to integrates these technologies and to improve the Environmental Information Systems (EIS). A key factor to improve the air quality monitoring is to share environmental data coming from different bodies (public and private companies) in a near real-time system, in order to take advantage of data from different sensor networks.

In this context, the present paper analyses some perspectives of low-cost (high density) monitoring network for a more direct control of the human health risk from atmospheric macro-pollutants.

1.2 MATERIALS AND METHODS

The work has been carried out in different steps, described in detail below.

As a first step, in the present paper a few micro-scale critical situations were selected pointing out the peak values of O_3 and NO_2, CO, that could be reached and the potential effects on health, in order to develop strategies and policies to improve the status of air quality and to comply with the National and European legislations.

Critical situations could be found generally:

- in the yard of kindergartens and schools (when an important road is present in the proximity);
- in street canyons (when the flux of traffic can be critical);
- in residential areas close to highways;
- in residential areas close to tunnels;
- in residential areas above trenched roads;
- in the proximity of large industrial plants;
- in summer in residential areas.

For these situations a selection of parameters to be investigated by wireless sensors has been made.

As a second step, low-cost sensors have been selected in order to check their viability to act as sentinels where the conventional approach of air quality monitoring cannot guarantee a high detail (that is in the cases analyzed in the first step).

As a third step, preliminary experiments were developed in the Torino city, near the central zone for five days in the summer period. The measurements were made outdoor, with the sensors put at 10 meters above the street level and then in an office.

As a fourth step, based also on this preliminary experience, an overall strategy has been developed for selected case studies. This step is sponsored by the Autonomous Province of Trento.

1.3 RESULTS AND DISCUSSION

In Table 1 the results of the analysis of the critical situations in terms of parameter selection are presented. The parameters have been filtered taking into account the availability of specific sensors in the sector. In particular, PM10 and similar parameters have been discarded as not yet suitable for a wireless sensor network.

TABLE 1: Selected parameters for potentially critical cases

Critical case	Selected parameters	Notes
Kindergartens	NO_2	CO could be added
Street canyon	NO_2	CO could be added
Highways proximity	NO_2	—
Tunnel proximity	NO_2	—
Trenched road proximity	NO_2	—
Industrial plants	CO, NO_2	Depending on process
Summertime	O_3	—

The parameter CO has not been considered in case of medium-high speed roads. The parameter NO has not been taken into account as not toxic. Large industrial plants could emit significantly either one of the two selected parameters, or both, or none of them: a preliminary analysis of the process is compulsory and the way of release into the atmosphere must be analyzed in details as well designed stacks could decrease the local impact to very low levels.

Then, the electrochemical low-cost sensors have been selected in order to act as sentinels in case of peak values of the parameters listed in Table 1.

The above listed sensors have been chosen taking into account their resolutions compared to the lowest peak value to be detected. For each parameter a group of low cost sensors will be adopted in order to generated data on an area.

For each low cost mini-network a high resolution sensor has been selected to be used as "mother" for a better interpretation of the generated data. The chosen NO_2, CO and O_3 sensors are thick low sensors.

FIGURE 1: The low-cost O$_3$ sensor

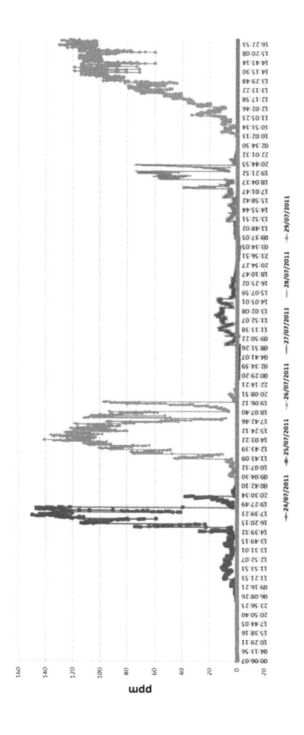

FIGURE 2: The overall preliminary low-cost O_3 sensor measurements

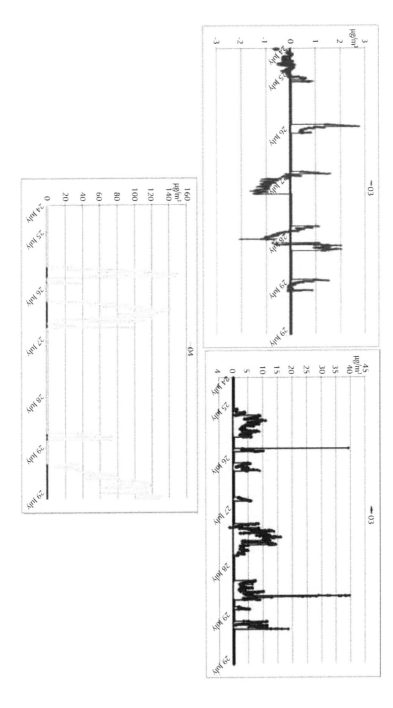

FIGURE 3: The preliminary low-cost O$_3$ sensor measurements

FIGURE 4: O$_3$ (and NO) measurements

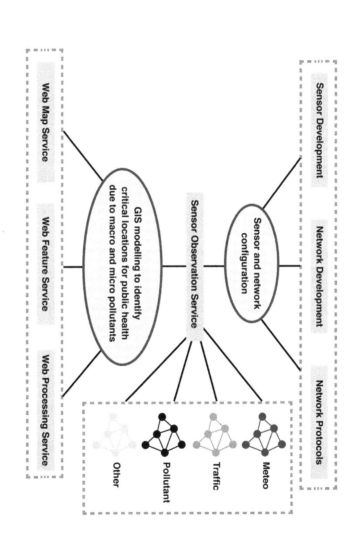

FIGURE 5: Overall view of the approach

TABLE 2: Selected case studies in Trentino

Case study	Mini-network parameters
Cement work where a different strategy of cocombustion is planned	NO_2 in winter in selected a residential area close to the plant
Paper mill plant a different strategy of co-generation is planned	NO_2 in winter in a selected residential area close to the plant
Sintering plant where a different limit of CO emissions could be set	CO in a selected residential area close to the plant
One residential area	O_3 in summer in a selected residential area
One tourist area	O_3 in summer in a selected tourist area

The sensors will be calibrated in the WSN Calibration Laboratory of Polytechnic University of Turin, DISMIC department. One of the low-cost sensors is presented in Figure 1.

Critical values of O_3 can be found in summer in residential areas. To this concern, a preliminary experience has been developed in Torino during summer 2011.

In Figure 2 the preliminary O_3 measurements made outdoor and indoor during the five days are presented, showing also the deep details. In Figure 3, instead, the O_3 measurements are presented taking into account the sensor position, respectively:

- in the laboratory with closed windows;
- in the laboratory with opened windows;
- in balcony at 10 meters above the street level (the highest peaks).

As a preliminary qualitative check of the sensor performance, the generated data have been compared with the ones of the closest fixed station of the local Environmental Protection Agency. In Figure 4 the data from an average day in summer detected by a fixed station of air quality of the ARPA Piemonte are reported. Afternoon peaks are detected in both cases. Of course a "cheek to cheek" calibration/validation have been planned in order to avoid distance effects.

Taking into account the preliminary results of the experimentation, the characteristics of the region where the field measurements have been planned (Trentino), the timing of the overall research proposed to the Autonomous Province of Trento (one year), the overall strategy presented in Table 2 has been planned.

The paper mill plant case study could be substituted by a test period along a highway. Additional modification of the strategy could be decided depending on optimization of timing and targets.

1.4 CONCLUSIONS

As regards to the use of low-cost sensors the sector seems to be ready for switching the air quality control strategy towards a direct health control. Low cost sensors for PM10 are not yet available but peaks of NO_2, CO and O_3 can be measured or detected.

The accurate detection of these peaks and the testing of the low cost sensor network in local conditions are essential to design proper sensor networks.

The future strategy, presented in Figure 5, of mixing different technologies like: low cost sensor networks with traditional sensor networks and other protocols promise to reach several advantages compared to traditional air quality monitoring systems:

- allows reaching higher spatial accuracy;
- reduces the redundancy of measures by different network systems;
- improves the localization of critical pollutants concentrations;
- reduces the costs improving data spatial resolution and quality;
- allows the creation of a real time alert system for dangerous pollutants.

The work in progress will be a significant reference experience. A progressive extension of the network is expected locally with private and public contributions. The same experience could be replicated easily thanks to the generation of guidelines for a correct approach wireless sensor based.

REFERENCES

1.   M. Ciolli, A. Cemin, D. Nave, "Modeling emission and dispersion of road traffic pollutant for the town of Trento", in Proceedings of Open Source Free Software GIS - GRASS users conference, University of Trento, Italy, 2002.

2. A. Cemin, M. Ciolli, M. Ragazzi, M. Zanoni, "Sistema integrato GIS-database per la gestione dei dati di traffico e produzione di mappe delle emissioni. Applicazione alla città di Trento". In Ingegneria Ambientale, vol. XXXIII/10, 2004, pp. 479–487.

3. V. Singh, C. Carnevale, G. Finzi, E. Pisoni, M. Volta, "A cokriging based approach to reconstruct air pollution maps, processing measurement station concentrations and deterministic model simulations", in Environmental Modelling & Software, vol. 26/6, 2011, pp. 778 – 786.

4. R. Beelen, M. Voogt, J. Duyzer, P. Zandveld, G. Hoek, 2010. "Comparison of the performances of land use regression modelling and dispersion modelling in estimating small-scale variations in long-term air pollution concentrations in a Dutch urban area". In Atmospheric Environment, vol. 44/36, 2010, pp. 4614 – 4621.

5. A. Mlter, S. Lindley, F. de Vocht, A. Simpson, R. Agius, "Modelling air pollution for epidemiologic research part i: A novel approach combining land use regression and air dispersion" in Science of The Total Environment, vol. 408/23, 2010, pp. 5862 – 5869.

6. M. Ragazzi, M. Grigoriu, E.C. Rada, E. Malloci, F. Natolino, "Risk assessment from combustion of sewage sludge treatment: three caste study comparison", in Proceedings of Recent Advanges in Risk Management, Assessment and Mitigation. RIMA'10, Bucharest, Romania, 2010, pp. 176–180.

7. Y. Liu, G. Cui, Z. Wang, Z. Zhang, "Large eddy simulation of wind field and pollutant dispersion in downtown Macao", in Atmospheric Environment, vol. 45/17, 2011, pp. 2849 – 2859.

8. M. Ciolli, M. De Franceschi, R. Rea, A. Vitti, D. Zardi, P. Zatelli, "Development and application of 2D and 3D GRASS modules for simulation of thermally driven slope winds", in Transactions in GIS, vol. 8/2, 2004, pp. 191–209.

9. M.S. Wong, J.E. Nichol, K.H. Lee, 2009. "Modelling of aerosol vertical profiles using gis and remote sensing", in Sensors, vol. 9/6, 2009, pp. 4380–4389.

10. M. Brini,"Chemical sensors suitability criteria for ubiquitous wireless sensor", in Proceedings of Networks Sensor Systems for Environmental Monitoring, London, U.K, 2010.

11. M. Brini, E.C. Rada, M. Ragazzi, L. Marmo, M. Chelodi, "Innovative pilot experience of nitrogen dioxide wireless sensor networking for human exposure assessment, in Proceedings of Towards the 2013 Revision of the Ambient Air Quality Directive - Issues and Solutions, London, UK, 2011.

PART II

WASTE INCINERATION

CHAPTER 2

A Review of Exposure Assessment Methods in Epidemiological Studies on Incinerators

MICHELE CORDIOLI, ANDREA RANZI, GIULIO A. DE LEO, AND PAOLO LAURIOLA

2.1 INTRODUCTION

Incineration is one of the most common technologies for waste disposal [1]. The number of incineration plants in Europe has been constantly rising in the last years, in the effort to manage and treat an ever-increasing waste production according to the EU directives and minimizing landfill disposal [2]. As waste incineration releases in the atmosphere chemicals that are potentially toxic [3], there is increasing public concern about the possible adverse effects on human health caused by this waste management technology [4, 5].

A Review of Exposure Assessment Methods in Epidemiological Studies on Incinerators. © Cordioli M, Ranzi A, De Leo GA, and Lauriola P. Journal of Environmental and Public Health **2013** *(2013). http://dx.doi.org/10.1155/2013/129470. Licensed under Creative Commons Attribution 3.0 Unported License, http://creativecommons.org/licenses/by/3.0/.*

The literature on health effects of waste incinerators is extensive and can be essentially classified into two groups: observational studies (i.e., epidemiological analyses) and simulation studies (i.e., health risk assessment). The first group includes studies that make use of a variety of statistical techniques to describe the potential relationship between the observed health status of the population and the exposure level from incinerators. The second group includes studies aimed at estimating the expected impact, in terms of health risk and/or number of sanitary cases, of a measured or simulated exposure to environmental contaminants [6–8].

Available epidemiological studies have been well reviewed in many published papers [9–11] and reports published by international agencies [12, 13]. However, the lack of a common framework for study designs makes the results of the different investigations on the health impacts hardly comparable and thus inconclusive. Poor exposure assessment is claimed as one of the main reasons of inconsistency of results in published studies [3, 9, 10, 13].

Exposure is generally defined as the contact between a stressor and a receptor and can be characterized either by direct (e.g., personal monitoring and biological markers) or indirect methods (e.g., environmental monitoring, modelling, and questionnaires) [14]. Although direct measures of exposure can be considered the best measures for assessing the effect of a specific substance on the target population, indirect measures of exposure (e.g., simulations of atmospheric dispersion) have greater utility for source emission assessment and control, since they are capable of linking population health to specific pollution emission sources [14]. These indirect methods have rapidly evolved in the last years [15], especially due to the increasing diffusion of the use of Geographical Information Systems (GIS) [16] and computer models to simulate atmospheric dispersion [17].

The aims of the present work were twofold: first, we wanted to investigate what methods and approaches are commonly used in the published literature to characterize exposure levels from waste incinerators; second, we wanted to assess, through a computer simulation study, how the classification of the expected exposure level may change as a function of the method used to estimate it.

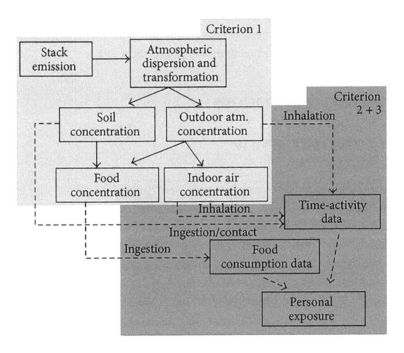

FIGURE 1: Conceptual model representing the principal impact pathways that determine exposure to atmospheric emissions from an incinerator. Contamination of drinking water is not represented.

The analysis was performed by using the literature database gathered within a project supported by the Emilia Romagna Region (North Italy) between 2007 and 2012 (MONITER Project) [18], to standardize environmental monitoring and health surveillance methods in areas characterized by the presence of incinerators and to evaluate the health status of populations around the eight incinerators of the region.

Although the focus of the present work was on waste incinerators, the results of our analysis can be extended to any point source of atmospheric pollution [19] or more generally to contaminated sites, where the presence of multiple sources has to be taken into proper account.

2.2 MATERIAL AND METHODS

2.2.1 LITERATURE REVIEW

We analyzed papers referenced in previously published reviews on incinerator health effects [9–13, 20] and, additionally, searched for further references on MEDLINE, PubMed, Scopus, and Google Scholar by using a number of keywords combinations (e.g., "epidemiology," "incinerator," "adverse effect," etc.). We focused our analysis only on observational epidemiological studies. Human biomonitoring [21, 22] and risk assessment studies [7, 8, 23] were not considered here. We excluded also studies on incinerator's workers [9] as the exposure pathway and levels can be completely different from those experienced by the population living around the incinerator plants.

The studies reviewed, rather than defining a relationship between environmental pollution and human health, aimed at evaluating the possible association with a specific industrial source of pollution (i.e., incinerators). The conceptual model for the emission-exposure pathways is sketched in Figure 1. Waste incineration epidemiological studies usually focus on gas stack emissions from the combustion process, while other possible sources of pollution (water discharges, ashes, smell emission, traffic, etc.) are not generally investigated [3]. After the emission from the incineration stack, pollutants dispersion in the atmosphere depends upon a number of physical and environmental variables such as stack height, wind speed and direction, temperature, and atmospheric stability. Some gases may undergo various chemical transformations, and part of the contaminants may eventually settle down on a variety of surfaces such as soil, vegetation, and water. Concentrations in the atmosphere and in soil may be either directly inhaled, ingested, or absorbed through dermal contacts or they can enter the agricultural food chain [24]. The actual exposure to potentially hazardous contaminants is thus determined by the time spent by various sectors of the population in different environments (outdoor, indoor at home, or at work) and could be due to inhalation, ingestion of contaminated water or food, and dermal contact with contaminated vectors (e.g., soil, water) [25]. Since incinerators are potential sources of persistent pollutants (e.g., dioxins, heavy metals, etc.) [3], ingestion can represent a relevant exposure pathway.

TABLE 1: Classification of exposure assessment methods.

Category	Description
Criterion 1: definition of exposure intensity	
1	Qualitative (e.g., presence/absence of the source/contamination in an area)
2	Distance from the source (e.g., linear distance)
3	Dispersion models (e.g., average annual atmospheric concentration)
Criterion 2: definition of population distribution	
1	Municipality/community/postcode sector
2	Census unit/full postcode
3	Exact residential address location
Criterion 3: temporal variability	
1	Time-invariable (i.e., fixed) exposure
2	Time-variable exposure (e.g., residential history and/or variability in emissions from the source)

Exposure to pollutants has conceptually at least three dimensions, namely, (i) the intensity of exposure, which depends among the other things upon the concentration level of contaminants in different media; (ii) space, as both population density and concentration of contaminants are spatially heterogeneous; (iii) time, which is the duration and variability of exposure, as this determines the total amount of contaminant that has been eventually ingested, inhaled, or absorbed through dermal contacts [14]. Exposure assessment reconstructs the relationship between receptors and locations and between locations and the presence and amount of a certain risk factor. Accordingly, we reviewed the selected literature focusing only on the approaches used to define the exposure level and classifying them on the basis of three criteria (Table 1):

1. the approach used to define the intensity of exposure to the emission source (3 categories);
2. the scale at which the spatial distribution of the exposed population was accounted for (3 categories);
3. whether temporal variability in exposure was considered or not in the published study (2 categories).

The combination of all categories can result in a total of 18 possible methods of exposure assessment and was hereafter referred to as "x.y.z," where x represents the method used to estimate expected intensity, y the method used to estimate population distribution, and z whether the exposure was variable or not in time. For example, a published study classified as "2.3.1" means that the exposure level was evaluated as a function of the distance from the source, population distribution in the territory was assessed by using exact residential address location, and exposure was fixed in time.

Exposure assessment methods were categorized only on the basis of the exposure variables actually used in the epidemiological model. As discussed afterward, some studies reported additional information (such as measured concentrations of pollutants in various media) useful to interpret or support exposure model outcomes, even though this information was not used in statistical calculations.

Another important element of the exposure assessment process is the control of confounding factors, that is, variables that may hide or enhance the measure of effect [26, 27]. These factors can be socioeconomic (e.g., people living in industrial areas near incinerators may be more deprived) or environmental (e.g., frequently incinerators are located in areas with high pollution from other industrial sources and traffic).

For each reviewed study we analysed also whether and how confounding factors were accounted for. Since evaluation of confounding factors can follow a variety of approaches, we decided not to include this aspect as a fourth criterion in our classification scheme. Nevertheless we thoroughly comment on the role of confounding factors as well as their importance in epidemiological studies in the discussion.

2.2.2 CASE STUDY: PARMA

To understand how the choice of one or another approach of Table 1 may ultimately affect the estimated exposure, we run a simulation case study based on real data from an epidemiological surveillance program for a new incinerator that is under construction in the city of Parma (Italy).

The data used to simulate the effect of alternative methods of exposure assessment were as follows:

1. location of the stack of the incinerator;
2. exact location of the address of residence for 31,019 people living around the incinerator (circle of 4 km of radius);
3. boundaries of the 2001 Italian census blocks for the area, as defined by the Italian National Institute of Statistics;
4. the results of an atmospheric dispersion model for PM_{10} emitted from the incinerator.

Geographic coordinates of addresses were provided by the local registry office. Atmospheric dispersion was simulated using the ADMS Urban model [28], a second generation quasi-Gaussian model already employed in other studies on health effects of incinerators [29–31]. Since the study area is located in a flat plane, this model was judged suitable to compute long-term average concentration and deposition [32].

We used PM_{10} as a tracer for the complex mix of pollutants emitted by the incinerator, after a test on various types of pollutant. The aim of the simulation was to determine a geographic gradient of exposure inside the study area: this spatial gradient is mainly determined by the incinerator's characteristics and atmospheric conditions, while it is only poorly dependent on the pollutant's properties.

We used five years of hourly meteorological data (2005–2010) from the nearest meteorological station (about 4 km from the plant) and source characteristics from the authorized project (i.e., stack height: 70 m; gas temperature: 150°C; PM_{10} emission flux: 231 mg s^{-1}) to calculate average hourly concentrations at ground level (ng m^{-3}) and average hourly deposition (ng m^{-2} h^{-1}) of PM_{10} over the period 2005–2010 on a regular 200 m receptor grid. Calculated concentrations were interpolated (using quadratic inverse distance weighting) to obtain continuous maps (Figure 3).

For each individual, we evaluated residential time-invariant exposure to the incinerator using the following methods:

1. distance between census block centroid and incinerator (CBDI, method 2.2.1);
2. distance between exact address location and incinerator (ADDI, method 2.3.1);
3. average concentration and deposition inside the census block of each address (CBCO and CBDE, method 3.2.1);
4. concentration and deposition at the address location (ADCO and ADDE, method 3.3.1).

We then contrasted the results of using alternative approaches for the assessment of the exposure level for each individual in the sample. Exposure variables were categorized in 5 classes (i.e., 1: lowest exposure, 5: highest exposure) using quintiles of each variable distribution. Thus, each exposure class contain approximately the same number of subjects. Only for address distance from the incinerator we defined also a second categorization using regular buffers, as done in the majority of published studies [33–35].

Concentration and deposition estimates based on dispersion models are affected by their own degree of uncertainty and should be possibly ground trued with field measurements and/or experiments. A previous validation study conducted in France [32] demonstrated that this kind of models provide a reliable proxy for incinerator exposure in simple terrain such as the area under study: we here assumed that simulated concentrations represent the closest estimate to the actual exposure.

Therefore we evaluated the degree of exposure misclassification using two-way tables and Cohen's kappa test of agreement [36, 37]. Cohen's kappa was calculated using quadratic weighting to assign less importance to misclassification between adjacent classes and higher importance to other misclassifications.

2.3 RESULTS

2.3.1 LITERATURE REVIEW

A total of 41 studies published between 1984 and January 2013 were identified by the literature search. Table S1 in Supplementary Material available online at http://dx.doi.org/10.1155/2013/129470 (Supplementary Information) reports the resulting categorization of exposure methods and other relevant information for each study. The column "covariates" lists the confounding factors that were evaluated in each study.

Figure 2 represents the evolution of methodologies in time, based on the year of publication. Methods on the -axis are sorted from the less precise to the best one.

With reference to the first classification criterion, that is, method used to assess exposure intensity, 19 studies (46%) used a measure of distance, both on a continuous scale and more commonly by defining concentric areas with arbitrary radius. In some cases [38–41] also wind direction was used to introduce some spatial anisotropy in exposure. Lee and Shy [42] used distance to define exposed communities but developed also a longitudinal study using daily PM_{10} measurement from fixed air monitors. One study [43] analysed spatial clustering of disease cases: since the analysis was based on the position of the community of residence, we classified this method as 2.1.1. One study [41] presented multiple assessment methods: presence/absence of the incinerator, distance from the plant, and an exposure index based on distance, wind direction, and time spent outdoor by people.

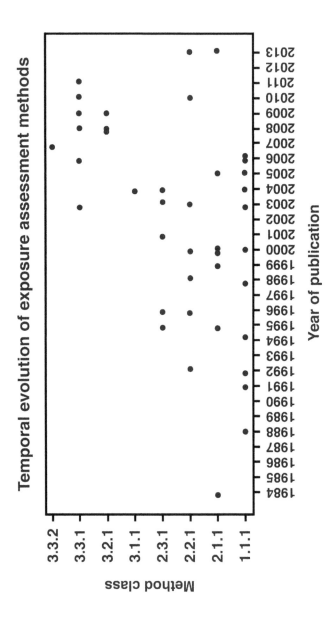

FIGURE 2: Temporal evolution of exposure assessment methods. Methods are classified according to Table 1 and sorted in the y-axis from the less precise to the best one.

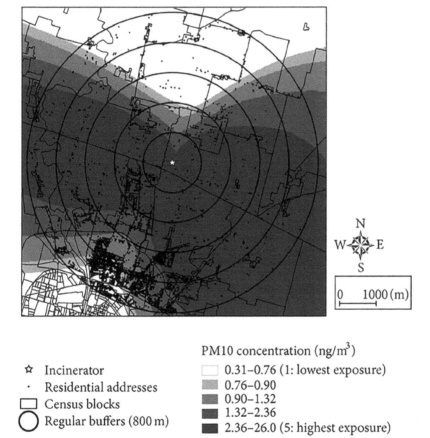

☆ Incinerator
· Residential addresses
▢ Census blocks
◯ Regular buffers (800 m)

PM10 concentration (ng/m^3)
▢ 0.31–0.76 (1: lowest exposure)
▨ 0.76–0.90
▨ 0.90–1.32
▨ 1.32–2.36
▨ 2.36–26.0 (5: highest exposure)

FIGURE 3: Representation of the area considered in the case study of Parma.

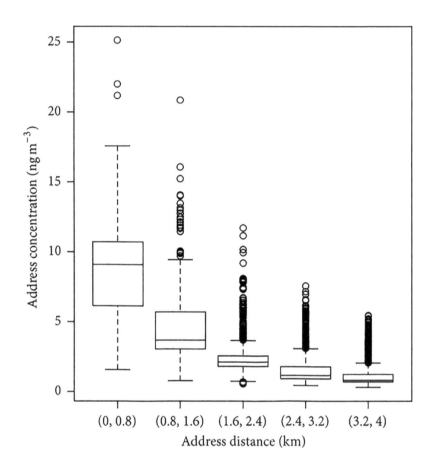

FIGURE 4: Results of exposure assessment by using different methodologies. (a) Variability of residential address concentration (ADCO) inside each regular 800 m buffer. Boxes represent the interquartile range (IQR), the horizontal line inside the box is the median value, and the whiskers extend to 1.5 times the IQR from the box. (b) Relationship between ground concentration (ADCO) and deposition (ADDE) at addresses location. The line represents the linear regression model. (c) Relationship between simulated concentrations evaluated at exact address (ADCO) and at census block level (CBCO). The line represents the 1 : 1 relationship. (d) Relationship between distance of the exact address ($ADDI_1$) and distance of the census block centroid (CBDI). The line represents the 1 : 1 relationship.

Eleven studies (27%) used atmospheric dispersion models to define population exposure. Generally models were used to estimate long-term average atmospheric concentrations at ground level, although one study used cumulated depositions [44]. Two studies [29, 45] used also heavy metals as indicator of exposure, while all the others used dioxins.

The remaining 11 studies (28%) used a qualitative definition of exposure to contrast the health status of communities/municipalities with and without incinerators. One study [46] developed quantitative indicators to classify municipalities, using emission inventories for dioxin from incinerators.

All the published studies used the residence as the place where exposure to atmospheric pollution occurs (criterion no.2). Nevertheless, different levels of detail were used in defining residence location. The majority of the papers (n = 19, 46%) considered the municipality or community of residence (e.g., postcode sector, school, hospital, etc.), 12 studies (29%) used the exact geographic coordinates of the address of residence, and 10 (24%) used the full postcode or census unit.

Finally, all the published literature, with one exception [47], defined exposure proxies that did not account for temporal variability in population spatial distribution and incinerators' emissions (criterion 3) that is, they considered the residence at the time of diagnosis, at enrolment, or the longest residence of the subject. Residential histories and changes in exposure intensity (e.g., as a consequence of changes in combustion and gas depuration technologies) were not accounted for in the other examined studies.

Overall, Figure 2 shows a trend of improvement in the quality of exposure assessment during the examined years, although three studies published after 2010 still used linear distance as the exposure proxy.

2.3.2 RESULTS OF THE SIMULATION CASE STUDY IN PARMA

Figure 3 reports the map of the census blocks around the incinerator under construction in Parma, its location (the star), the location of the sample of resident people used in the present study (small black dots), the expected PM_{10} concentration as simulated with the ADMS model, and the regular, 800 m wide, circular buffers around the emission source. Figure 4 contrasts the results of alternative approaches to assess exposure level in terms of intensity (simulated concentration versus distance from the emission sources) and accuracy in residence location.

Table 2 shows Cohen's kappa indices of agreement between concentration maps and other exposure assessment methods. The table reports also the share of individuals over the 31,019 samples assigned to the same class of exposure, the share of individuals classified in an adjacent exposure class, and that of individuals classified into two or more classes apart. High kappa values are encountered when concentrations and depositions are considered, while comparison between concentration and distance approaches gave worst results when distance is categorized on regular concentric circles.

2.4 DISCUSSION

2.4.1 EVALUATION OF EXPOSURE INTENSITY (CRITERION 1)

The majority of the papers reviewed in the present study appear to suffer from poor exposure characterization. A relevant part of these papers (28%) used qualitative definitions of exposure (e.g., presence/absence of the source or anecdotic presence of pollution). These methods cannot account for the complexity of impact pathways described in Figure 1 nor for the heterogeneity in the exposure level that is normally expected as a consequence of the uneven distribution of the resident population and of the anisotropic dispersal of pollutants in the atmosphere. For instance, in the simulation case study we ran in Parma, the use of method 1.1.1 (presence of the incinerator in the municipality) would not allow us to discriminate between different levels of exposure and, therefore, all the 30,019 people in our sample (as well as the remaining 158,660 inhabitants of Parma) would be all classified as highly exposed, which would probably not be the case.

Epidemiological analyses carried out on a significant number of municipalities still represent a valuable instrument for public health tracking since they can evidence disease clusters in some regions that must be studied further. Even though any departure of disease incidence in large communities from background levels has to be taken very seriously, it is very difficult to use this type of evidence to infer about the role of specific emission sources (i.e., an incinerator), as many other potential confounding factors might exert a significant effect, particularly in highly urbanized areas. Moreover, the risk of false positive and, to a greater extent, false negative results, common to all exposure assessment methods, can be exacerbated when epidemiological data are averaged out on a vast territory with large internal differences in the exposure levels, as in method 1.1.1.

TABLE 2: Evaluation of the agreement between concentration maps and other exposure assessment methods. Quadratic weighted Cohen's kappa and percentages of subjects classified in the same exposure class or in different classes.

Comparison exposure	Weighted kappa[a]	Matching subjects	Misclassification in adjacent categories	Misclassification in >1 class apart
ADCO versus ADDE	0.91	69.6%	29.3%	1.1%
ADCO versus CBCO	0.97	89.2%	10.5%	0.3%
ADCO versus CBDE	0.90	70.0%	27.8%	2.2%
ADCO versus ADDI1	0.61	38.9%	45.1%	16.0%
ADCO versus CBDI	0.60	40.2%	44.5%	15.3%
ADCO versus ADDI2	0.35	25.4%	39.8%	34.8%

ADCO: address concentration (quintiles), ADDE: address deposition (quintiles), CBCO: census block concentration (quintiles), CBDE: census block deposition (quintiles), $ADDI_1$: address distance (quintiles), $ADDI_2$: address distance (regular 800 m buffers), CBDI: distance between census block centroid and incinerator. [a]all kappa with $p < 0.001$.

Almost half of the studies used distance to measure exposure. This is certainly a substantial improvement with respect to just an absence/presence evaluation, as contamination from an atmospheric emission source (e.g., air, soil, and locally produced food) is generally expected to decrease with distance. However, the assumption of isotropy in atmospheric dispersion of contaminants could lead to remarkable errors in exposure assessment. Many features of the emission source (e.g., stack height, gas flow temperature and velocity, and pollutant concentration) and of the local environment (e.g., local meteorology, topography, and land use) determine where and how far stack emissions disperse and how ultimately enter different environmental compartments.

In our simulation study carried out for the Parma incinerator, the distance method assigns the same exposure level to people resident in the northern and eastern parts of the territory around the emission stack, even though simulations showed that concentrations are expected to be higher along the east-west direction than to the north-south one (Figure 3).

Because of the anisotropic dispersion of pollutants in the atmosphere, the expected PM_{10} concentrations at the residence address vary wildly inside each 800 m wide buffer around the incinerator (Figure 4(a)). Consequently, the use of distance from the emission source as a proxy of actual concentrations could cause a high degree of misclassification (Table 2).

The use of well-tuned atmospheric dispersion models allows a substantial improvement in the estimation of exposure level, especially if carried out along with a fine scale estimation of the spatial distribution of the vulnerable population. Anyway, atmospheric pollution models are themselves affected by a considerable level of uncertainty [48] depending upon assumptions on actual atmospheric conditions, reconstruction of wind fields, and type of dispersion processes, including the possibility of simulating chemical transformation which are known to be highly relevant for the formation of tropospheric ozone and secondary fine particulate matter.

A significant number of the published papers analysed in the present study provided only a limited information on the atmospheric model: generally there was no discussion about the type of model used, the type and source of meteorological data, model adequacy to represent complex morphological natural or urban landscape and/or wind calms, and the assumptions made about pollutant's emission rates and physical-chemical properties.

Only few studies explicitly acknowledged limitations in the modelling approach used. For example, instead of adopting a different dispersion model as suggested by the same authors in a previous study [32], in Viel et al.'s [49] a part of the study area was excluded because dispersion model results were judged unreliable in that area. Another study [50] used maps of ground level concentrations estimated on the basis of emissions and meteorological data, but no dispersion model was cited. Almost all the studies used dioxins as an impact indicator: dioxins represent a family of 210 congeners, each one with different physical-chemical characteristics: no study clearly explained how these chemicals were treated in the model (e.g., using 2,3,7,8-TCDD congener properties). Moreover, some studies did not report a clear definition even of the most basic variables used to measure exposure, that is; averaging time for concentrations [31, 45, 51] or distinction between concentrations and depositions to ground [31]. As shown in our case study ground level atmospheric PM_{10} concentrations and depositions from a point source have very similar patterns with some significant departure, nevertheless the choice of one or the other measure of exposure should be at least discussed, related to the main route of exposure considered. All these pieces of information are important to judge the quality of the exposure assessment process, its uncertainties, and to allow comparability and reproducibility of methods.

Regardless of how detailed, accurate and advanced the model to simulate atmospheric dispersion is, it is still only a part of the impact pathways described in Figure 1. All the studies implicitly assumed that inhalation represents the principal exposure pathway, while no published literature measured or modelled the possible exposure through ingestion of contaminated food or contact with contaminated soil.

No study used measured levels of pollution in different media (e.g., atmosphere, soil, and food) as the exposure variable in the epidemiological model, except for one work [42] that used also measured 24 h average PM_{10} concentrations in each community as a predictor for pulmonary function, although there were no differences in average levels between communities defined a priori as exposed and not exposed. Many studies presented information on measured levels of pollution [43, 52–54], but these data were not included in the statistical model. This is not surprising, as it is very difficult to discriminate the contribution of single-point sources to the observed concentrations levels. The latter, in fact, invariably depend the contribution of several other confounding emission sources [55, 56], especially if they are located in urbanized areas with intense traffic or industrial activities. Thus, indirect measures of exposure obtained through modelling represent a valid alternative useful to identify the possible role of a specific emission source.

2.4.2 EVALUATION OF RECEPTOR'S EXPOSURE (CRITERIA 2 AND 3)

The actual exposure of an individual to the pollutants emitted by an incinerator may occur in different environments and last a variable amount of time. All published studies used the residence as the place where exposure to atmospheric pollution occurs (criterion 2). Notably, one study [57] considered also the location of workplace of studied subjects.

Residence location can be determined with various degrees of precision. The majority of revised studies (48%) used community level to determine residence location (i.e., town, municipality, postcode sector, and school). In this way the same exposure level is assigned to large groups of population, but this assumption was rarely discussed and no measures of exposure variability inside groups were reported. Thus, it was impossible to evaluate the degree of ecological bias [58] that is, how well the variation in risk between groups with different average exposure applies to the variation in risk between individuals.

Some studies used census block or full postcode for determining residence position. The dimension of these blocks may vary greatly depending on the location: normally these blocks are smaller in populated areas but may become very large in other rural zones. Moreover, no information was generally given about blocks extension, and it was difficult to compare very different blocks types like Small Area Health Statistic Unit (SAHSU) [35], UK census postcode system [59], or UK Lower Layer Super Output Areas (LSOA) [60]. In our case study census blocks had an average area of $0.4\,\mathrm{km^2}$ (min: $968.4\,\mathrm{m^2}$; max: $6.3\,\mathrm{km^2}$) and contained on average 26 addresses (min: 1; max: 130): both address distances and concentrations vary widely inside some census blocks (Figures 4(c) and 4(d)). This was true especially for more exposed areas, since the incinerator is located in a less densely populated area with large census blocks. This aspect could lead to different degree of errors in exposure assignment, that increase with the level of pollutant or proximity to the incinerator.

The most precise way to locate residences is to address geocoding: this procedure assigns a couple of geographic coordinates to each address. Errors in address positioning depend on the quality of the database used but is generally in the order of tens to hundreds meters [61, 62], thus small in comparison with the use of census blocks or full postcode.

In future studies maximum disaggregation of data, to maximise information and minimize potentially differential ecological biases [63], is thus recommended.

The use of residence as exposure location is a very common assumption in environmental epidemiology since it is easily derived and there is evidence that people normally spend a great part of their time inside their residences, for example, on average 69% [64] and 80% [65]. Nevertheless, home location may not well represent total exposure because people may experience shorter but more intense exposures outside home, and residence is a proxy only for inhalation exposure and does not account for indirect pathways [66] (Figure 1). Although this technique has well-known limitations, it is often the only method available, particularly for large populations or for reconstructing historical exposures.

Temporal variability in exposure is an issue rarely explored in the reviewed studies. Temporal variability may result both from changes in source emissions over time or from residential mobility of the population and may be a cause of incorrect exposure assignment [67, 68]. Only one published study [47] explicitly accounts for historical exposure variability by reconstructing residential histories and evolution of dioxin emissions from the sources considered. However the exposure indicator chosen (i.e., the average exposure over time) may introduce some bias: since emissions from the sources considered were progressively reduced starting from the 1990s, the average exposure value decreases with the increase of exposure duration. A better indicator could have been cumulative exposure, that is, the sum of the annual exposure concentration over the exposure duration. One study [29] considered the modification of incinerator emissions over time indirectly, without considering changes in the final statistical model, but evaluating how the morphology of fallout maps was similar in time.

Although difficult to achieve because of data unavailability, especially for studies on old incinerators, in future studies efforts should be developed in reconstructing residential histories and variability in incinerator's emission over time, at least as a sensitivity analysis for the model.

2.4.3 EXPOSURE MISCLASSIFICATION
AND CONFOUNDING FACTORS

Almost all papers used categorical definitions of exposure (i.e., exposure classes). One issue rarely discussed is the rationale behind the choice of cut-off values used to classify continuous variables. In the absence of toxicological reference values for this type of exposure, in our case study we used a criterion expected to make the results of the statistical analysis more stable and reliable, that is, having roughly the same number of exposed individuals in each class. In reviewed studies a priori cutoffs of exposure were generally chosen without an explicit justification [33–35, 51].

When categorical exposure variables are measured with error, they are said to be misclassified. Misclassification can be differential or nondifferential with respect to disease status of an individual person [26], the latter being more probable in reviewed studies and generally leading to risk estimations biased toward the null. Nevertheless, in presence of more than two exposure categories, non-differential misclassification can move estimates of risk away from null and disrupt exposure-response trends [69].

Our case study showed that:

1. for exposure measures based on distance a relevant part of the population may be classified in the wrong exposure category (assuming that dispersion model better represents real exposure), with relevant percentages of subjects moving by more than one category;
2. the use of census blocks to identify the residence may introduce a certain degree of differential misclassification since the error is higher in more exposed areas and lower for less exposed.

Both these factors may bias risk estimates away from the null or modify exposure-response trends.

Sometimes, the degree of error in exposure assessment can be evaluated with a validation study [70], that is, comparing modelled exposure with "gold-standard" measurement of exposure collected for a random subsample of the population, such as direct measurement of individual exposure. In practice, since no such gold standard is generally available, we recommend researchers to conduct sensitivity analyses on exposure assessment [71] and discuss the magnitude of error that may be present in their data.

Another issue that is only partially dealt with in reviewed literature is confounding. Confounding occurs when a risk factor different from the exposure variable under study causes bias in the estimation of association between exposure and disease, due to its differential distribution in exposed and non exposed groups [72]. Various confounding factors may affect a study on incinerators' health effects, that is, socioeconomic differences (e.g., poverty, occupation), personal lifestyles (e.g., alcohol, smoke), and presence of other sources of pollution.

Many reviewed studies did not account for any confounder in the epidemiological model [33, 47, 59, 73–77]. Some studies collected information about personal lifestyles or socio-economic status directly through questionnaires [38–40, 51, 78, 79]. Unfortunately the use of questionnaires and surveys is unfeasible for large populations; thus a large part of the studies did not consider personal lifestyles but included socio-economic indicators (e.g., deprivation indexes) evaluated at municipality/census block of residence [29, 30, 35, 44, 45, 49, 80, 81]. These indexes are generally constructed based on census statistics.

Of particular concern is the general lack of information about environmental confounding. Many of the pathologies under study have been associated with various atmospheric pollutants (e.g., PM_{10}, NO_x, etc.) or specific anthropogenic sources (e.g., road traffic, industrial emissions). Often, waste incinerators are located inside industrial areas or near other major sources of pollution. In our case study, for example, the incinerator is located inside the industrial area of Parma, at about 200 m from a national highway that crosses the study area east-west (i.e., the prevalent wind directions). As a result, most exposed subjects, as identified by the dispersion model, were also more exposed to other sources of pollution. It will be difficult to correctly identify the possible health effect of this incinerator, unless we have some information about the difference in population exposure to other sources between the exposed and nonexposed groups. Only few studies included information about environmental confounders. Biggeri et al. [79] used measured particulate depositions from the nearest monitoring station, Cordier et al. [45] used proxies for the presence of industrial activities and road traffic at community level, and two studies

[31, 44] used proxies for traffic and industrial pollution at census block level. Notably, one recent study [29] used atmospheric dispersion models to estimate pollution concentrations at the address of residence from other local sources of atmospheric pollution (road traffic, industrial plants, and heating). This represents a notable improvement since the confounding factor was evaluated with the same spatial resolution as exposure to the incinerator.

As the quantitative contribution of well-managed modern incinerators to total pollution levels in a study area and to baseline health risks is expected to be low, we suggest to draw a careful attention to other local sources of pollution and to implement multisite studies on large populations where feasible.

2.5 CONCLUSIONS

We reviewed 41 articles from the literature with the main aim of retrieving information for the definition of an exposure assessment protocol to be used in a large study on health effects of pollution due to incinerators (MONITER project).

Overall, our analysis showed a trend of improvement in exposure assessment quality over time, with a massive use of dispersion models in exposure assessment after year 2003.

Nevertheless, the lack of a common framework for exposure assessment is demonstrated by the use of a variety of methods, also in recent papers, with different quality of epidemiological findings and difficulties in comparisons of results.

In most of the selected studies the characterization of exposure can be significantly improved by using more detailed data for population residency and better simulation models. Recent development of informative systems and high availability of environmental and demographic data suggest the use of dispersion models of pollutants emitted from a source, combined with precise methods of geographic localizations of people under study, as the state of the art method to assess exposure of population in epidemiological studies. Considerations about residential mobility, temporal variations in pollution emissions, latency period of investigated diseases, and treatment of environmental and sociodemographic confounders can improve exposure assessment accuracy.

All these aspects of exposure assessment are particularly relevant as most of environmental conflicts usually arise from the evaluation of the contribution of the various pollution sources to the overall contamination.

REFERENCES

3. Eurostat, "Eurostat statistical books. Environmental statistics and accounts in Europe 2010. 2010 Edition," European Union, 2010.
4. D. Saner, Y. B. Blumer, D. J. Lang, and A. Koehler, "Scenarios for the implementation of EU waste legislation at national level and their consequences for emissions from municipal waste incineration," Resources, Conservation and Recycling, vol. 57, pp. 67–77, 2011.
5. M. F. Reis, "Solid waste incinerators: health impacts," in Encyclopaedia of Environmental Health, pp. 162–217, 2011.
6. M. Allsopp, P. Costner, and P. Johnston, "Incineration and human health: state of knowledge of the impacts of waste incinerators on human health (executive summary)," Environmental Science and Pollution Research, vol. 8, no. 2, pp. 141–145, 2001.
7. British Society for Ecological Medicine, The Health Effects of Waste Incinerators 4th Report of the British Cociety for Ecological Medicine, 2nd edition, 2008.
8. F. Forastiere, C. Badaloni, K. De Hoogh et al., "Health impact assessment of waste management facilities in three European countries," Environmental Health, vol. 10, no. 1, article 53, 2011.
9. R. J. Roberts and M. Chen, "Waste incineration—how big is the health risk? A quantitative method to allow comparison with other health risks," Journal of Public Health, vol. 28, no. 3, pp. 261–266, 2006.
10. M. Cordioli, S. Vincenzi, and G. A. Delleo, "Effects of heat recovery for district heating on waste incineration health impact: a simulation study in Northern Italy," Science of the Total Environment, vol. 444, pp. 369–380, 2013.
11. D. Porta, S. Milani, A. I. Lazzarino, C. A. Perucci, and F. Forastiere, "Systematic review of epidemiological studies on health effects associated with management of solid waste," Environmental Health, vol. 8, no. 1, article 60, 2009.
12. M. Franchini, M. Rial, E. Buiatti, and F. Bianchi, "Health effects of exposure to waste incinerator emissions: a review of epidemiological studies," Annali dell'Istituto Superiore di Sanita, vol. 40, no. 1, pp. 101–115, 2004.
13. S.-W. Hu and C. M. Shy, "Health effects of waste incineration: a review of epidemiologic studies," Journal of the Air and Waste Management Association, vol. 51, no. 7, pp. 1100–1109, 2001.
14. U. K. Defra, Review of Environmental and Health Effects of Waste Management: Municipal Solid Waste and Similar Wastes, 2004.

15. WHO Europe, Population Health and Waste Management: Scientific Data and Policy Options, Copenhagen, Denmark, 2007.

16. Committee oh Human and Environmental Exposure in the 21st entury, Exposure Science in the 21st Century. A Vision and a StrAtegy, National Acedemy of Sciences, 2012.

17. M. Nieuwenhuijsen, D. Paustenbach, and R. Duarte-Davidson, "New developments in exposure assessment: the impact on the practice of health risk assessment and epidemiological studies," Environment International, vol. 32, no. 8, pp. 996–1009, 2006.

18. J. R. Nuckols, M. H. Ward, and L. Jarup, "Using geographic information systems for exposure assessment in environmental epidemiology studies," Environmental Health Perspectives, vol. 112, no. 9, pp. 1007–1015, 2004.

19. B. Zou, J. G. Wilson, F. B. Zhan, and Y. Zeng, "Air pollution exposure assessment methods utilized in epidemiological studies," Journal of Environmental Monitoring, vol. 11, no. 3, pp. 475–490, 2009.

20. Emilia Romagna Region, "The MONITER project," 2012, http://www.arpa.emr.it/moniter/index.asp.

21. A. Kibble and R. Harrison, "Point sources of air pollution," Occupational Medicine, vol. 55, no. 6, pp. 425–431, 2005.

22. L. Giusti, "A review of waste management practices and their impact on human health," Waste Management, vol. 29, no. 8, pp. 2227–2239, 2009.

23. H. L. Chen, H. J. Su, P. C. Liao, C. H. Chen, and C. C. Lee, "Serum PCDD/F concentration distribution in residents living in the vicinity of an incinerator and its association with predicted ambient dioxin exposure," Chemosphere, vol. 54, no. 10, pp. 1421–1429, 2004.

24. M. F. Reis, C. Sampaio, P. Aguiar, J. Maurício Melim, J. Pereira Miguel, and O. Päpke, "Biomonitoring of PCDD/Fs in populations living near portuguese solid waste incinerators: levels in human milk," Chemosphere, vol. 67, no. 9, pp. S231–S237, 2007.

25. M. Schuhmacher, J. L. Domingo, and J. Garreta, "Pollutants emitted by a cement plant: health risks for the population living in the neighborhood," Environmental Research, vol. 95, no. 2, pp. 198–206, 2004.

26. M. Schuhmacher, K. C. Jones, and J. L. Domingo, "Air-vegetation transfer of PCDD/PCDFs: an assessment of field data and implications for modeling," Environmental Pollution, vol. 142, no. 1, pp. 143–150, 2006.

27. EPA, "Human health risk assessment protocol for hazardous waste combustion facilities," Tech. Rep. EPA530-R-05-006, 2005.

28. A. Blair, P. Stewart, J. H. Lubin, and F. Forastiere, "Methodological issues regarding confounding and exposure misclassification in epidemiological studies of occupational exposures," American Journal of Industrial Medicine, vol. 50, no. 3, pp. 199–207, 2007.

29. L. Sheppard, R. T. Burnett, A. A. Szpiro et al., "Confounding and exposure measurement error in air pollution epidemiology," Air Quality, Atmosphere and Health, vol. 5, no. 2, pp. 203–216, 2011.

30. Cambridge Environmental Research Consultants Ltd, ADMS URban technical specification, 2001.

31. A. Ranzi, V. Fano, L. Erspamer, P. Lauriola, C. A. Perucci, and F. Forastiere, "Mortality and morbidity among people living close to incinerators: a cohort study based on dispersion modeling for exposure assessment," Environmental Health, vol. 10, no. 1, article 22, 2011.

32. S. Cordier, A. Lehébel, E. Amar et al., "Maternal residence near municipal waste incinerators and the risk of urinary tract birth defects," Occupational and Environmental Medicine, vol. 67, no. 7, pp. 493–499, 2010.

33. J.-F. Viel, C. Daniau, S. Goria et al., "Risk for non Hodgkin's lymphoma in the vicinity of French municipal solid waste incinerators," Environmental Health, vol. 7, article 51, 2008.

34. N. Floret, J.-F. Viel, E. Lucot et al., "Dispersion modeling as a dioxin exposure indicator in the vicinity of a municipal solid waste incinerator: a validation study," Environmental Science and Technology, vol. 40, no. 7, pp. 2149–2155, 2006.

35. P. Comba, V. Ascoli, S. Belli et al., "Risk of soft tissue sarcomas and residence in the neighbourhood of an incinerator of industrial wastes," Occupational and Environmental Medicine, vol. 60, no. 9, pp. 680–683, 2003.

36. P. Michelozzi, D. Fusco, F. Forastiere, C. Ancona, V. Dell'Orco, and C. A. Perucci, "Small area study of mortality among people living near multiple sources of air pollution," Occupational and Environmental Medicine, vol. 55, no. 9, pp. 611–615, 1998.

37. P. Elliott, G. Shaddick, I. Kleinschmidt et al., "Cancer incidence near municipal solid waste incinerators in Great Britain," British Journal of Cancer, vol. 73, no. 5, pp. 702–710, 1996.

38. S. Peters, R. Vermeulen, A. Cassidy et al., "Comparison of exposure assessment methods for occupational carcinogens in a multi-centre lung cancer case—control study," Occupational and Environmental Medicine, vol. 68, no. 2, pp. 148–153, 2011.

39. C. Eriksson, M. E. Nilsson, D. Stenkvist, T. Bellander, and G. Pershagen, "Residential traffic noise exposure assessment: application and evaluation of European Environmental Noise Directive maps," Journal of Exposure Science and Environmental Epidemiology, 2012.

40. F. Barbone, M. Bovenzi, A. Biggeri, C. Lagazio, F. Cavallieri, and G. Stanta, "Comparison of epidemiologic methods in a case-control study of lung cancer and air pollution in Trieste, Italy," Epidemiologia e Prevenzione, vol. 19, no. 63, pp. 193–205, 1995.

41. C. M. Shy, D. Degnan, D. I. Fox et al., "Do waste incinerators induce adverse respiratory effects? An air quality and epidemiological study of six communities," Environmental Health Perspectives, vol. 103, no. 7-8, pp. 714–724, 1995.

42. A. K. Mohan, D. Degnan, C. E. Feigley et al., "Comparison of respiratory symptoms among community residents near waste disposal incinerators," International Journal of Environmental Health Research, vol. 10, no. 1, pp. 63–75, 2000.

43. S.-W. Hu, M. Hazucha, and C. M. Shy, "Waste incineration and pulmonary function: an epidemiologic study of six communities," Journal of the Air and Waste Management Association, vol. 51, no. 8, pp. 1185–1194, 2001.

44. J.-T. Lee and C. M. Shy, "Respiratory function as measured by peak expiratory flow rate and PM10: six communities study," Journal of Exposure Analysis and Environmental Epidemiology, vol. 9, no. 4, pp. 293–299, 1999.
45. J.-F. Viel, P. Arveux, J. Baverel, and J.-Y. Cahn, "Soft-tissue sarcoma and non-Hodgkin's lymphoma clusters around a municipal solid waste incinerator with high dioxin emission levels," American Journal of Epidemiology, vol. 152, no. 1, pp. 13–19, 2000.
46. S. Goria, C. Daniau, P. De Crouy-Chanel et al., "Risk of cancer in the vicinity of municipal solid waste incinerators: importance of using a flexible modelling strategy," International Journal of Health Geographics, vol. 8, no. 1, article 31, 2009.
47. S. Cordier, C. Chevrier, E. Robert-Gnansia, C. Lorente, P. Brula, and M. Hours, "Risk of congenital anomalies in the vicinity of municipal solid waste incinerators," Occupational and Environmental Medicine, vol. 61, no. 1, pp. 8–15, 2004.
48. Y. Fukuda, K. Nakamura, and T. Takano, "Dioxins released from incineration plants and mortality from major diseases: an analysis of statistical data by municipalities," Journal of Medical and Dental Sciences, vol. 50, no. 4, pp. 249–255, 2003.
49. P. Zambon, P. Ricci, E. Bovo et al., "Sarcoma risk and dioxin emissions from incinerators and industrial plants: a population-based case-control study (Italy)," Environmental Health, vol. 6, article 19, 2007.
50. K. S. Rao, "Uncertainty analysis in atmospheric dispersion modeling," Pure and Applied Geophysics, vol. 162, no. 10, pp. 1893–1917, 2005.
51. J.-F. Viel, M.-C. Clément, M. Hägi, S. Grandjean, B. Challier, and A. Danzon, "Dioxin emissions from a municipal solid waste incinerator and risk of invasive breast cancer: a population-based case-control study with GIS-derived exposure," International Journal of Health Geographics, vol. 7, article 4, 2008.
52. R. Tessari, C. Canova, F. Canal et al., "Environmental pollution from dioxins and soft tissue sarcomas in the population of Venice and Mestre: an example of the use of current electronic information sources," Epidemiologia e Prevenzione, vol. 30, no. 3, pp. 191–198, 2006.
53. N. Floret, F. Mauny, B. Challier, P. Arveux, J.-Y. Cahn, and J.-F. Viel, "Dioxin emissions from a solid waste incinerator and risk of non-Hodgkin lymphoma," Epidemiology, vol. 14, no. 4, pp. 392–398, 2003.
54. S. Parodi, R. Baldi, C. Benco et al., "Lung cancer mortality in a district of La Spezia (Italy) exposed to air pollution from industrial plants," Tumori, vol. 90, no. 2, pp. 181–185, 2004.
55. T. Tango, T. Fujita, T. Tanihata et al., "Risk of adverse reproductive outcomes associated with proximity to municipal solid waste incinerators with high dioxin emission levels in Japan," Journal of Epidemiology, vol. 14, no. 3, pp. 83–93, 2004.
56. E. J. Gray, J. K. Peat, C. M. Mellis, J. Harrington, and A. J. Woolcock, "Asthma severity and morbidity in a population sample of Sydney school children: part I—prevalence and effect of air pollutants in coastal regions," Australian and New Zealand Journal of Medicine, vol. 24, no. 2, pp. 168–175, 1994.
57. S. Caserini, S. Cernuschi, M. Giugliano, M. Grosso, G. Lonati, and P. Mattaini, "Air and soil dioxin levels at three sites in Italy in proximity to MSW incineration plants," Chemosphere, vol. 54, no. 9, pp. 1279–1287, 2004.

58. M. Nadal, M. C. Agramunt, M. Schuhmacher, and J. L. Domingo, "PCDD/PCDF congener profiles in soil and herbage samples collected in the vicinity of a municipal waste incinerator before and after pronounced reductions of PCDD/PCDF emissions from the facility," Chemosphere, vol. 49, no. 2, pp. 153–159, 2002.

59. M. Vinceti, C. Malagoli, S. Teggi et al., "Adverse pregnancy outcomes in a population exposed to the emissions of a municipal waste incinerator," Science of the Total Environment, vol. 407, no. 1, pp. 116–121, 2008.

60. G. Shaddick, D. Lee, and J. Wakefield, "Ecological bias in studies of the short-term effects of air pollution on health," International Journal of Applied Earth Observation and Geoinformation, vol. 22, pp. 65–74, 2013.

61. F. L. R. Williams, A. B. Lawson, and O. L. Lloyd, "Low sex ratios of births in areas at risk from air pollution from incinerators, as shown by geographical analysis and 3-dimensional mapping," International Journal of Epidemiology, vol. 21, no. 2, pp. 311–319, 1992.

62. N. F. Reeve, T. R. Fanshawe, T. J. Keegan, A. G. Stewart, and P. J. Diggle, "Spatial analysis of health effects of large industrial incinerators in England, 1998–2008: a study using matched case-control areas," BMJ Open, vol. 3, Article ID e001847, 2013.

63. P. A. Zandbergen, "Geocoding quality and implications for spatial analysis," Geography Compass, vol. 3, no. 2, pp. 647–680, 2009.

64. D. T. Duncan, M. C. Castro, J. C. Blossom, G. G. Bennett, and L. G. G. G. Steven, "Evaluation of the positional difference between two common geocoding methods," Geospatial Health, vol. 5, no. 2, pp. 265–273, 2011.

65. P. Diggle and P. Elliott, "Disease risk near point sources: statistical issues for analyses using individual or spatially aggregated data," Journal of Epidemiology and Community Health, vol. 49, no. 2, pp. S20–S27, 1995.

66. N. E. Klepeis, W. C. Nelson, W. R. Ott et al., "The National Human Activity Pattern Survey (NHAPS): a resource for assessing exposure to environmental pollutants," Journal of Exposure Analysis and Environmental Epidemiology, vol. 11, no. 3, pp. 231–252, 2001.

67. X. Wu, D. H. Bennett, K. Lee, D. L. Cassady, B. Ritz, and I. Hertz-Picciotto, "Longitudinal variability of time-location/activity patterns of population at different ages: a longitudinal study in California," Environmental Health, vol. 10, no. 1, article 80, 2011.

68. Y.-L. Huang and S. Batterman, "Residence location as a measure of environmental exposure: a review of air pollution epidemiology studies," Journal of Exposure Analysis and Environmental Epidemiology, vol. 10, no. 1, pp. 66–85, 2000.

69. M. A. Canfield, T. A. Ramadhani, P. H. Langlois, and D. K. Waller, "Residential mobility patterns and exposure misclassification in epidemiologic studies of birth defects," Journal of Exposure Science and Environmental Epidemiology, vol. 16, no. 6, pp. 538–543, 2006.

70. J. R. Meliker, M. J. Slotnick, G. A. AvRuskin, A. Kaufmann, G. M. Jacquez, and J. O. Nriagu, "Improving exposure assessment in environmental epidemiology: application of spatio-temporal visualization tools," Journal of Geographical Systems, vol. 7, no. 1, pp. 49–66, 2005.

71. M. Dosemeci, S. Wacholder, and J. H. Lubin, "Does nondifferential misclassification of exposure always bias a true effect toward the null value?" American Journal of Epidemiology, vol. 132, no. 4, pp. 746–748, 1990.

72. C. A. Holcroft and D. Spiegelman, "Design of validation studies for estimating the odds ratio of exposure- disease relationships when exposure is misclassified," Biometrics, vol. 55, no. 4, pp. 1193–1201, 1999.

73. D. Spiegelman, "Approaches to uncertainty in exposure assessment in environmental epidemiology," Annual Review of Public Health, vol. 31, pp. 149–163, 2010.

74. R. McNamee, "Confounding and confounders," Occupational and Environmental Medicine, vol. 60, no. 3, pp. 227–234, 2003.

75. H. Rydhstroem, "No obvious spatial clustering of twin births in sweden between 1973 and 1990," Environmental Research, vol. 76, no. 1, pp. 27–31, 1998.

76. G. W. Ten Tusscher, G. A. Stam, and J. G. Koppe, "Open chemical combustions resulting in a local increased incidence of orofacial clefts," Chemosphere, vol. 40, no. 9-11, pp. 1263–1270, 2000.

77. A. Biggeri and D. Catelan, "Mortality for non-Hodgkin lymphoma and soft-tissue sarcoma in the surrounding area of an urban waste incinerator. Campi Bisenzio (Tuscany, Italy) 1981–2001," Epidemiologia e Prevenzione, vol. 29, no. 3-4, pp. 156–159, 2005.

78. A. Biggeri and D. Catelan, "Mortalità per linfomi non Hodgkin nei comuni della Regione Toscana dove sono stati attivi inceneritori di rifiuti solidi urbani nel periodo 1970–1989," Epidemiologia & Prevenzione, vol. 1, pp. 14–15, 2006.

79. F. Bianchi and F. Minichilli, "Mortalità per linfomi non Hodgkin nel periodo 1981–2001 in comuni italiani con inceneritori di rifiuti solidi urbani," Epidemiologia & Prevenzione, vol. 2, pp. 80–81, 2006.

80. T.-R. Hsiue, S.-S. Lee, and H.-I. Chen, "Effects of air pollution resulting from wire reclamation incineration on pulmonary function in children," Chest, vol. 100, no. 3, pp. 698–702, 1991.

81. A. Biggeri, F. Barbone, C. Lagazio, M. Bovenzi, and G. Stanta, "Air pollution and lung cancer in Trieste, Italy: spatial analysis of risk as a function of distance from sources," Environmental Health Perspectives, vol. 104, no. 7, pp. 750–754, 1996.

82. P. Elliott, M. Hills, J. Beresford et al., "Incidence of cancers of the larynx and lung near incinerators of waste solvents and oils in Great Britain," The Lancet, vol. 339, no. 8797, pp. 854–858, 1992.

83. M. Federico, M. Pirani, I. Rashid, N. Caranci, and C. Cirilli, "Cancer incidence in people with residential exposure to a municipal waste incinerator: an ecological study in Modena (Italy), 1991–2005," Waste Management, vol. 30, no. 7, pp. 1362–1370, 2010.

Management of Atmospheric Pollutants from Waste Incineration Processes: The Case of Bozen

MARCO RAGAZZI, WERNER TIRLER, GIULIO ANGELUCCI, DINO ZARDI, AND ELENA CRISTINA RADA

3.1 INTRODUCTION

In the sector of municipal solid waste (MSW) management, incineration options have undergone significant improvements in recent years, as demonstrated in many life cycle analysis studies and research (Consonni et al., 2005; Damgaard et al., 2010; Fruergaard and Astrup, 2011; Larsen and Astrup, 2011; Marculescu, 2012). However, at least for the sector of MSW, incineration still remains the major contributor of polychlorinated dibenzo-p-dioxin and polychlorinated dibenzofuran (PCDD/F) emissions (Rada et al., 2006). In a wider context, metallurgy and uncontrolled burning are the most relevant sources in countries where environmental controls are more carefully operated (Antunes et al., 2012; Lemieux et al., 2004). As

Management of Atmospheric Pollutants from Waste Incineration Processes: The Case of Bozen. © *Ragazzi M, Tirler W, Angelucci G, Zardi D, and Rada EC.* Waste Management and Research *31,3 (2013). DOI: 10.1177/0734242X12472707. Reprinted with permission from the authors.*

the method of release into the atmosphere may significantly change the role of a MSW plant in terms of PCDD/F immissions (Rada et al., 2011), the characterization of the performance of a MSW incinerator should take into account both the assessment of stack emissions and the measurements of pollutants at ground level. In most countries flue gas dispersion modeling takes place in each prevalent weather type, with its concomitant wind direction and speed, and simulates this effect. In this respect a recent article pointed out how the contribution of a modern incinerator, assessed by modeling, can be very low compared with the values that the local monitoring stations can typically measure (Ragazzi and Rada, 2012).

The European stack emission limit for municipal waste incineration plants is 0.1 ng_{TEQ} Nm^{-3}, and similar limits are discussed for other industrial plants. PCDD/F stack gas sampling at stationary sources is usually performed by manual sampling according to Ente Nazionale Italiano di UNIficazione (UNI) norms (UNI, 1999a,1999b,1999c). This allows sampling times between 6 and 8 hours. Only 2–3 characterizations per year are needed to comply with the regulation. Public concern often focuses the PCDD/F emissions during the remaining time. Indeed, the use of a long-term sampling system can enable sampling for a week or even longer (Kahr and Steiner, 2002a; Reinmann, 2002). Thanks to the larger sampling volume detection, statistical variations can be reduced (Tirler et al., 2003).

In addition to PCDD/F, MSW incineration, like many other plants including combustion processes, may be a potential source of other atmospheric pollutants, especially fine particles. As in the case of asbestos, with its long latency (in the order of decades), there are many legitimate concerns about the unknown human health consequences from nano-materials and, specifically, ultra-fine particles from specific sources, such as incinerators (Brunner et al., 2006). The need to understand the role of the incinerator of Bozen, Italy, in terms of local contribution to the concentration of ultra-fine particles, motivated the development of specific research activity. The contents of the present article refer to some results from this multi-year research activity, which were taken into account during the local decision process aimed at implementing a new plant (presently under construction) to substitute the one which is the subject of the present case-study. The latter had to be substituted because of concerns about the structure, as its cement structure was built in 1994. However, the need for a new plant was considered as an opportunity to update the adopted technology for reducing pollutant emissions and improving energy generation.

The plant is located in an area renowned for its natural heritage, environmental quality, intensive mountain agriculture and beautiful landscapes. Furthermore, more than 100,000 people live in the urban and suburban areas of the city of Bozen, close to the plant. The incinerator is designed on two parallel lines, with a total treatment capacity of up to 400 t d^{-1} of MSW. The combustion takes place on a roller-type grate with a secondary combustion chamber; energy is recovered through a water tube boiler and a steam turbine, providing electricity and heat for district heating. Flue gas cleaning is performed by a fabric filter and a two-stage wet scrubber, in line with a final selective catalytic reduction unit for nitrogen oxides (NOx) and trace organics conversion.

A fabric filter aims to remove most of the particulate matter from the off-gas. The wet scrubber plays a secondary role for this parameter as its main purpose is the removal of acid gases.

Only residual municipal solid waste is treated at the plant. As a consequence, the chlorine content in the input is always far from critical values for hydrogen chloride and PCDD/F generation. Temperatures ranging between 900°C and 1000°C typically occur in the combustion chamber. The oxygen content in the off-gas is kept higher than 7% in order to favor a highly efficient combustion.

3.2 METHODS

3.2.1 EMISSION CONTROL OF PCDD/F

PCDD/F sampling, clean-up and quantification of the incineration plant of Bozen were performed in accordance with the present European Standard Protocols using $^{13}C_{13}$-labeled standards (UNI 1999a, 1999b, 1999c). The flue gas is released through a 50-m high stack, with an inner diameter of 2.5 m. Since 2003, when an automatic sampling system, suitable for monitoring PCDD/F emissions during the complete operation time of the plant, was installed at this incinerator, two probes (2500 mm and 1300 mm long, respectively) allow representative samples to be taken in the relatively large stack.

Short-term (8 h) flue gas sampling was performed by the filter-cooler method and conducted with an automatic, continuous adjusting isokinetic sampling system. Extraction of filter media was done by soxhlet with toluene and extraction of aqueous liquid by liquid/liquid extraction with dicloromethane. An automatic two-column system was used for chromatographic sample clean-up. Pre-packed Teflon® disposable columns containing multilayer silica and alumina were utilized. A long-term sampling system is a mechanically-engineered system that resembles the manual sampling approach for PCDD/F emitted from stationary sources. A small quantity of the flue gas was sampled continuously on a sampling cartridge. Depending on the sampling time—which ranged from 6 hours to 6 weeks—variously representative measurement results were obtained.

3.2.2 IMMISSION CONTROL OF PCDD/F

The meteorological characteristics linked to release processes concerning the Bozen municipal waste incinerator were examined. The target area lies in the Adige Valley in the Alps, running approximately in a North–South direction. A peculiar consequence of the above conditions is the frequent occurrence of along-valley winds. In particular, the latter typically blow up-valley during daytime and down-valley during night-time in the warm seasons under fair weather conditions (de Franceschi and Zardi, 2009; de Franceschi et al., 2002, 2009; Rampanelli et al., 2004). Accordingly the main deposit sites were identified south and north of the plant respectively, on the basis of a previous numerical modelling assessment (DICA, 2001). The interested reader will find additional information (e.g. dilution factors) in a recently published paper (Ragazzi & Rada, 2012). The first monitoring campaign, carried out in August and September 2006, used two directional air samplers placed to the north and south of the incinerator, each with two cartridges and collecting samples with the wind coming from the north and south respectively.

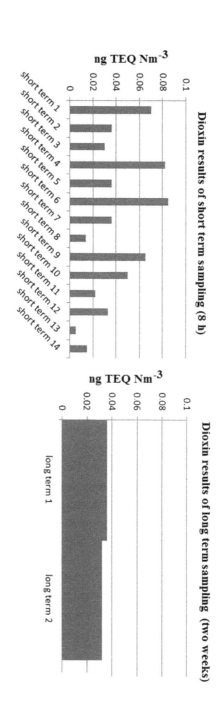

FIGURE 1: Results of short-term and of continuous, long-term (two weeks) sampling at the stack; results express the average concentration value over the sampling period.

To obtain more information, a second sampling campaign (December 2006–January 2007) with three directional air samplers was performed. The third sampler was installed in the centre of Bozen, in an area not influenced directly by the incinerator, collecting on two sampling cartridges when the wind came from north or south respectively. The other two samplers still worked in the same place north and south of the incineration plant. These samplers were able to work with three cartridges to collect the air, with the third cartridge also able to work under calm wind conditions. The parameters analyzed were PCDD/F, polycyclic aromatic hydrocarbons and, for the samplers north and south of the incinerator, also particulate matter $(PM)_{10}$, $PM_{2.5}$ and PM_1.

The adopted test standards were EPA 1613 for PCDD/F (EPA, 1994), gravimetric assessment for PMs and gas chromatographic–isotope dilution coupled with high-resolution mass spectrometry for benzo(a)pyrene (BaP).

3.2.3 ULTRA-FINE PARTICLE MEASUREMENTS

To measure the emissions of particles with an aerosol spectrometer directly on the stack is usually not possible. Stack gas can contain many compounds and also a relatively high amount of water. Therefore, condensation can occur when the gas is cooled down during sampling. Condensation will trap most of the particles and bias the obtained results. Condensation of liquid will also damage the measurement devices. So dilution is necessary in the case of MSW incineration plants. In the present case a modified continuous PCDD/F sampling system from monitoring systems was used (Kahr and Steiner, 2002b). The sampler works by the dilution method described in EN-1948 Part 1 (MCERTS, 2010). The sampling equipment permits dilution of flue gas while maintaining automatic isokinetic sampling. In the present case the hot stack gas (160°C) was diluted with pre-cleaned air at a ratio of 1:7. The continuous PCDD/F sampling system was coupled directly to aerosol spectrometer for the determination of fine and ultra-fine particles. The instrument used to detect fine dust was an aerosol spectrometer manufactured by Grimm (model 1.108). The dilution of the flue gas avoids condensation within the aerosol spectrometer.

The instrument makes it possible to determine the number of particles, as well as the particle mass. The aerosol spectrometer classifies the particles size in 15 different size channels from 0.25 up to 20 μm. The obtained size distribution spectrum permits qualitative consideration about the origin of PM. Sampling for ultra-fine particles was performed in the same way as for fine particles. For the detection of ultra-fine particles a condensation particle counter from Grimm, Ainring, Germany (model CPC 5403) was used, coupled with a Vienna-type differential mobility analyzer 55706 operating within a range from 5.5 and 350 nm. The granulometric classification of the ultra-fine particles is based on analysis of their electrical mobility.

3.3 RESULTS AND DISCUSSION

Results are presented starting from PCDD/F emission and immission data; PM emission and immission measurements complete this section. In Figure 1 examples of short- and long-term sampling are reported. The short-term PCDD/F concentrations measured at the stack are included between 10% and 80% of the emission limit—the long term average being about a third of it.

Concerning immissions, PCDD/F concentrations determined in the first campaign showed values ≤ 20 fg_{TEQ} m^{-3}. With northerly winds the immissions resulted in 6 fg_{TEQ} m^{-3} at the northern sampling point (Figure 2, site 3) and 20 fg_{TEQ} m^{-3} at the southern sampling point (Figure 2, site 5). With southerly winds the immissions resulted in 11 fg_{TEQ} m^{-3} at the northern sampling point and 15 fg_{TEQ} m^{-3} at the southern sampling point. These measurements showed values of PCDD/F in the air comparable with the ones of domestic/rural areas in summer.

Results from the second sampling campaign showed that the influence of the incineration plant on the PCDD/F concentration in the air of Bozen is simply not visible, confirming the results of previous preliminary research on the influence of this plant (Caserini et al., 2004): in winter (22 December 2006–18 January 2007) PCDD/F air concentrations ranged from 79 to 90 fg_{TEQ} m^{-3} at the Northern sampling point and from 52 to 84 fg_{TEQ} m^{-3} at the southern sampling point, whilst at the third site (Figure

2, site 6), not exposed to the main contribution of the plant, the values ranged from 155 to 259 fg_{TEQ} m^{-3}. The highest value was measured at the north sampling site under northerly winds. When this air was advected by the wind to the southern sampling site, a lower concentration was measured. The same situation occurred under southerly winds. The northern sampling site showed similar results. Relatively high concentrations were found under calm wind conditions. The above facts provide clear evidence that domestic heating and road traffic are the main sources of persistent organic pollutants (POPs) in the air of the areas analyzed. Concerning the above results, previous modeling confirmed the low significance of the plant contribution in terms of PCDD/F concentrations (DICA, 2001). Indeed, the two areas of highest impact are reported in Figure 2, where the results of the cited modeling are presented for PCDD/F. The maximum values are significantly lower than the ones measured in the ambient air in the present case study: even 2–3 orders of magnitude lower. In Figure 2 six points are marked in order to point out significant locations for unconventional measurements: 1—incineration stack; 2—incineration area close to a highway; 3—northern rural area; 4—suburban area; 5—southern rural area close to the highway; and 6—urban area.

Active air sampling presents a powerful tool in monitoring wind transport of POPs (Tirler et al., 2007a). The measurement results reported in Table 1 show that the waste incineration plant of Bozen is not a significant source of PCDD/F, PAHs or fine particles (the sampling site is the number 5 in Figure 2). Indeed, the concentrations of PCDD/F in ambient air are low compared with other case studies from the European Union (EU) (Vilavert et al., 2012) and non-EU countries (Bakoglu et al., 2005; Xu et al., 2009). In areas influenced by incineration with regulation not as stringent as the one of EU, the ambient air concentration can reach even some pg_{I-TEQ} m^{-3} (Xu et al., 2009). Regarding the distribution of organic pollutants, approximately 90% could be found in the PM_1 fraction. This fraction presents, in the cited study, only 70% of the PM_{10} fraction. PM_1 has a higher specific surface where organic pollutants can easily be bound on the surface. Data from Table 1 show that when the wind blows from the south the PM concentrations at the site show values higher than the ones when the wind blows from the north; taking into account the location of the incinerator, northern with respect to the site, the conclusion is that the plant is not a significant source of PM.

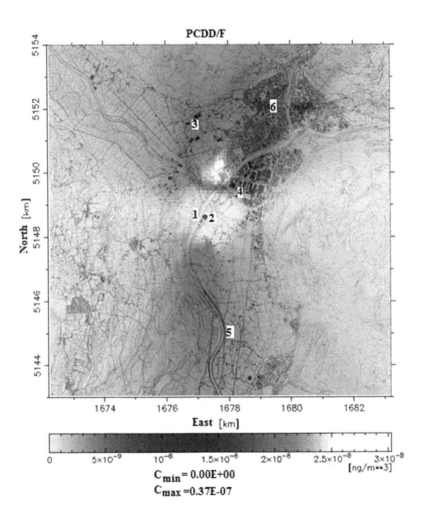

FIGURE 2: Medium annual concentration at ground level for polychlorinated dibenzo-p-dioxin and polychlorinated dibenzofuran (1: plant location; 2–6: sampling sites).

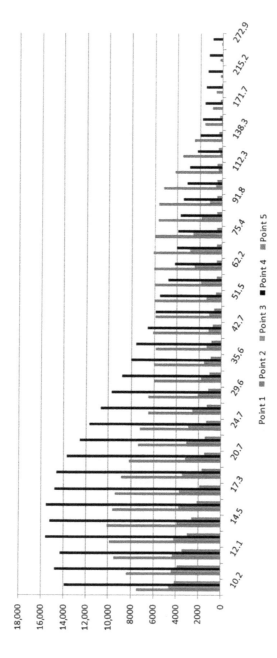

FIGURE 3: Ultrafine particles at the stack (sampling point 1) and in 4 additional points (sites 2–5).

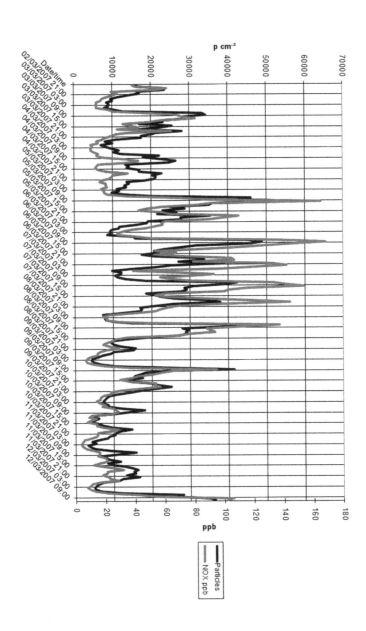

FIGURE 4: Time trend for ultra-fine particles (from 10 nm to 300 nm) and NOx in the ambient air in Bolzano.

TABLE 1: Polychlorinated dibenzo-p-dioxin and polychlorinated dibenzofuran (PCDD/F) and benzo(a)pyrene (BaP) in the different particular matter (PM) fractions in the southern sampling site.

	PM10		PM2.5		PM1	
	PCDD/F		PCDD/F		PCDD/F	
	$\mu g\ m^{-3}$	$fg_{TEQ}\ m^{-3}$	$\mu g\ m^{-3}$	$fg_{TEQ}\ m^{-3}$	$\mu g\ m^{-3}$	$fg_{TEQ}\ m^{-3}$
Wind from north	44	59	38	56	33	53
Wind from south	78	52	70	50	63	47
Calm of wind	45	84	39	82	34	
	77					
		BaP		BaP		BaP
	$\mu g\ m^{-3}$	$ng\ m^{-3}$	$\mu g\ m^{-3}$	$ng\ m^{-3}$	$\mu g\ m^{-3}$	$ng\ m^{-3}$
Wind from north	44	2.2	38	2.1	33	1.9
Wind from south	78	2.6	70	2.5	63	2.3
Calm of wind	45	3.1	39	3.0	34	2.8

Measured data showed very low fine dust concentrations (0.25–20 μm) in the flue gas; indeed, the cleaning system of the plant showed a high efficiency in removing fine particles (Tirler et al., 2007b). The levels of some classes were lower than in the air outside of the plant, which is also dependent on the presence of a highway close to the plant.

In the case of ultra-fine particles, the measured flue gas concentrations (Figure 3) were lower than the ones in the area of the plant from where the combustion air is taken. This means that the off-gas treatment line has a high efficiency in removing them. This characteristics were analyzed and quantified recently (Buonanno et al., 2012).

Figure 3 also illustrates some other interesting aspects:

- the presence of ultra-fine particles in the northern rural area is relatively low—at this site the proximity of the area of highest impact to the incinerator does not influence the values; indeed, its contribution is expected to be low (the dilution of ultra-fine particles emitted at the stack makes the role of the plant negligible);
- the contribution of the highway is visible if compared with the rural area, but seems not to be the only significant source; indeed, in the rural site close to the highway (Figure 2, site 5) the ultra-fine concentration is lower than in the urban and suburban sites;

- data in Figure 3 are reported as number of particles in a reference volume; indeed, in case of ultra-fine particles the mass concentration is not a significant parameter—from this point of view the peaks of ultra-fine particles in the flue gas and in the ambient air are different: the granulometric class showing the highest value in the flue gas is less fine than the one of the peak in the ambient air characterization.

Data concerning the number of particles were also correlated with data taken from the air quality monitoring stations of Bozen center (Figure 2, site 6). There was good correlation as far as the parameters directly linked to vehicle traffic (benzene, toluene, carbon monoxide, nitrogen dioxide, nitric oxide, NOx and number of vehicles) were concerned. An example is shown in Figure 4. The contribution of NOx from the plant in the urban area of Bozen (Figure 2, site 6) was demonstrated to be negligible (Ragazzi et al., 2012), thus the NOx dynamics can be attributed to the traffic. The similarity of the trends of NOx and ultra-fine particles is clear. Thus, ultra-fine particles in the urban area depends on the traffic dynamics.

3.4 CONCLUSIONS

From the results described herein, it can be stated that the ultra-fine dust measurements for the town of Bozen could be correlated with traffic: the site close to the incineration stack, not influenced by its emissions, shows high values of ultra-fine particles in an area far from the town, but close to a highway (this highway crosses the town); however, the lower values measured in another site, a rural area close to the highway, demonstrate that the highway is not the only significant source of ultra-fine particles. The fluctuations of NOx concentration in an urban site show the same trend of the fluctuations of the number of ultra-fine particles, with peaks that are synchronized with the peaks of local traffic. Other sources of ultra-fine particles seem to be present, as demonstrated by the high values measured in a suburban area far from the highway and main roads, but additional studies should be developed (in particular, the role of domestic combustion of wood could be investigated). Finally, the waste incineration plant of Bozen is not a significant source of fine and ultra-fine particles, nor for PCDD/F. The availability of a detailed pollutant diffusion modeling (in 2001), of a long-term PCDD/F emission characterization (since 2003) and of ultra-fine concentration values (in 2006–2007) allowed the local authorities to manage successfully the pathway of the proposal of the new incineration plant presently under construction.

REFERENCES

1. Antunes P, Viana P, Vinhas T, Rivera J and Gaspar EMSM (2012) Emission pro-files of polychlorinated dibenzodioxins, polychlorinated dibenzofurans (PCDD/Fs), dioxin-like PCBs and hexachlorobenzene (HCB) from secondary metallurgy indus-tries in Portugal. Chemosphere 88: 1332–1339.
2. Bakoglu M, Karademir A and Durmusoglu E (2005) Evaluation of PCDD/F levels in ambient air and soils and estimation of deposition rates in Kocaeli, Turkey. Che-mosphere 59: 1373–1385.
3. Brunner TJ, Wick P, Manser P, Spohn P, Grass RN, Limbach LK, et al. (2006). In vi-tro cytotoxicity of oxide nanoparticles: comparison to asbestos, silica, and the effect of particle solubility. Environmental Science & Technology 40: 4374–4381.
4. Buonanno G., Scungio M, Stabile L and Tirler W (2012) Ultrafine particle emission from incinerators: The role of the fabric filter. Journal of the Air and Waste Manage-ment Association 62: 103–111.
5. Caserini S, Cernuschi S, Giugliano M, Grosso M, Lonati G and Mattaini P (2004) Air and soil dioxin levels at three sites in Italy in proximity to MSW incineration plants. Chemosphere 54: 1279–1287.
6. Consonni S, Giugliano M and Grosso M (2005) Alternative strategies for energy recovery from municipal solid waste: Part B: Emission and cost estimates. Waste Management 25: 137–148.
7. Damgaard A, Riber C, Fruergaard T, Hulgaard T and Christensen TH (2010) Life-cycle-assessment of the historical development of air pollution control and energy recovery in waste incineration. Waste Management 30: 1244–1250.
8. de Franceschi M and Zardi D (2009) Study of wintertime high pollution episodes during the Brenner-South ALPNAP measurement campaign. Meteorology and At-mospheric Physics 103: 237–250.
9. de Franceschi M, Rampanelli G and Zardi D (2002) Further investigations of the Ora del Garda valley wind. In: Proceedings of the 10th AMS Conference on Mountain Meteorology and MAP Meeting 2002, 17–21 June 2002, Park City, UT, USA, pp. 30–33. Boston, MA: American Meteorological Society.
10. de Franceschi M, Zardi D, Tagliazucca M and Tampieri F (2009) Analysis of second order moments in the surface layer turbulence in an Alpine valley. Quarterly Journal of the Royal Meteorological Society 135: 1750–1765.
11. Dipartimento di Ingegneria CIvile e Ambientale (Department of Civil and Environ-mental Engineering) (eds) (2001) Study of the emissions, atmospheric diffusion and deposition of the pollutants emitted from the Bozen incinerator. Research report,

Civil and Environmental Engineering Department Research, Faculty of Engineering, University of Trento, Italy. September

12. EPA (1994) Method 1613: tetra- through octa-chlorinated dioxins and furans by isotope dilution HRGC/HRMS.

13. Fruergaard T and Astrup T (2011) Optimal utilization of waste-to-energy in an LCA perspective. Waste Management 31: 572–582.

14. Kahr G and Steiner T (2002a) Obtaining dioxin values with low uncertainty using automatic long-term-sampling equipment and data evaluation. Organohalogen Compounds 59: 97–100.

15. Kahr G and Steiner T (2002b) Application of continuous dioxin monitoring technique according the European standard at high dust levels At a brick production plant and calculation of the annual mass flow. Organohalogen Compounds 59: 101–102.

16. Larsen AW and Astrup T (2011) CO2 emission factors for waste incineration: Influence from source separation of recyclable materials. Waste Management 31: 1597–1605.

17. Lemieux PM, Lutes CC and Santoianni DA (2004) Emissions of organic air toxics from open burning: a comprehensive review. Progress in Energy and Combustion Science 30: 1–32.

18. Marculescu C (2012) Comparative analysis on waste to energy conversion chains using thermal-chemical processes. Energy Procedia 18: 604–611.

19. MCERTS (2010) Method implementation document (MID 1948) Part 2: Extraction and clean-up of PCDDs/PCDFs. London: Environment Agency.

20. Rada EC, Ragazzi M, Panaitescu V and Apostol T (2006) The role of bio-mechanical treatments of waste in the dioxin emission inventories. Chemosphere 62: 404–410.

21. Rada EC, Ragazzi M, Zardi D, Laiti L and Ferrari A (2011) PCDD/F environmental impact from municipal solid waste bio-drying plant. Chemosphere 84: 289–295.

22. Ragazzi M and Rada EC (2012) Multi-step approach for comparing the local air pollution contributions of conventional and innovative MSW thermo-chemical treatments. Chemosphere 89: 694–701.

23. Rampanelli G, Zardi D and Rotunno R (2004). Mechanisms of up-valley winds. Journal of Atmospheric Sciences 61; 3097–3111.

24. Reinmann J (2002) Results of one year continuous monitoring of the PCDD/F emissions of waste incinerators in the Wallon region of Belgium with Ames. Organohalogen Compounds 59: 77–80.

25. Tirler W, Angelucci G, Bedin K, Voto G, Donegà M and Minach L (2007a) Active sampling and analysis of dioxins and polyaromatic hydrocarbons bound to fine particles in the vicinity of a municipal solid waste incinerator. Organohalogen Compounds 69: 2268–2271.

26. Tirler W, Angelucci G, Bedin K and Verdi L (2007b) Fine particles, ultra-fine and nano-particles in emission of a municipal solid waste incineration plant. Organohalogen Compounds 69: 1030–1033.

27. Tirler W, Donegà M, Voto G and Kahr G (2003) Quick evaluation of long term monitoring samples and the uncertainty of the results. Organohalogen Compounds 60: 509–512.

28. UNI (1999a) Stationary source emissions – determination of the mass concentration of PCDDs/PCDFs – Sampling. UNI EN 1948–1.

29. UNI (1999b) Stationary source emissions – determination of the mass concentration of PCDDs/PCDFs – extraction and clean-up. UNI EN 1948–2.

30. UNI (1999c) Stationary source emissions – determination of the mass concentration of PCDDs/PCDFs – identification and quantification. UNI EN1948-3.Vilavert L, Nadal M, Schuhmacher M and Domingo JL (2012) Long-term monitoring of dioxins and furans near a municipal solid waste incinerator: Human health risks. Waste Management and Research 30: 908–916 .

31. Xu MX, Yan JH, Lu SY, Li XD, Chen T, Ni MJ, et al. (2009) Concentrations, profiles, and sources of atmospheric PCDD/Fs near a municipal solid waste incinerator in Eastern China. Environmental Science & Technology 43:1023–1029.

CHAPTER 4

Comparative Assessment of Particulate Air Pollution Exposure from Municipal Solid Waste Incinerator Emissions

DANIELLE C. ASHWORTH, GARY W. FULLER,
MIREILLE B. TOLEDANO, ANNA FONT, PAUL ELLIOTT,
ANNA L. HANSELL, AND KEES DE HOOGH

4.1 INTRODUCTION

Incineration is being increasingly used as a waste management option in the United Kingdom (UK). This is in response to EU legislation restricting the amount of waste disposed of in landfills [1]. Up until the 1990s incineration in the UK was largely uncontrolled. Legislation pertaining to all incinerators in the UK, the EU Waste Incineration Directive (WID) (2000/76/EC), came into operation for new incinerators in 2002 and older ones in 2005. This has set strict limits on emissions into the air [2]; none-

*Comparative Assessment of Particulate Air Pollution Exposure from Municipal Solid Waste Incinerator Emissions. © Ashworth DC, Fuller GW, Toledano MB, Font A, Elliott P, Hansell AL, and de Hoogh K. Journal of Environmental and Public Health **2013** (2013), http://dx.doi.org/10.1155/2013/560342.*

theless, there remains public concern and scientific uncertainties about possible health risks from pollutants emitted from incinerators.

European waste legislation uses the Waste Hierarchy Framework to guide the use of different waste management options, prioritising the more environmental desirable and sustainable options. Incineration falls above disposal of waste in landfills within this framework but is not as desirable as recycling and composting, reuse, and prevention [3]. Municipal solid waste incinerators (MSWIs) burn waste assembled by collection authorities [4], at high temperatures, reducing the volume of waste, eliminating pathogens and are capable of recovering energy from the waste [5].

To date a number of epidemiological studies have investigated the relationship between incineration and health [4–12], with most focused on its association with risk of cancer and more recently, the risk of adverse birth outcomes [8, 12–24]. The UK Committee on Carcinogenicity of Chemicals in Food, Consumer Products and the Environment released a statement about MSWIs and cancer in 2000 (updated in 2009), stating that, "…any potential risk of cancer due to residency near to municipal solid waste incinerators was exceedingly low and probably not measureable by the most modern epidemiological techniques" [6, 7]. This was supported by the UK Health Protection Agency's' statement in 2009 "…While it is not possible to rule out adverse health effects from modern, well regulated municipal incinerators with complete certainty, any potential damage to the health of those living close-by is likely to be very small, if detectable" [25].However, the evidence base investigating this issue remains limited and most existing studies suffer from incomplete information on potential confounders, lack of statistical power, and poor exposure assessment.

Exposure assessment is often referred to as the "Achilles heel" of environmental epidemiology [26, 27]. Inaccurate and imprecise exposure estimates, leading to exposure misclassification, can create biases in health risk estimates. In many environmental epidemiology studies, exposure misclassification is unrelated to the health outcome, termed nondifferential exposure misclassification, which would be expected to bias observed effect estimates towards the null [28]. Accurate exposure assessment is particularly important for studies trying to detect/exclude small excesses in risk in relation to environmental exposures [29], such as due to incineration, in order to enable true risks, if present, to be detected.

The methods used to assess exposure to an environmental source, such as an incinerator, range in design and complexity, from simple proxy methods to detailed individual level measures of exposure. Simple proxy methods, such as distance to the incinerator, assume a linear decrease in exposure with distance from source but benefit from the ease of implementation and the limited data and resources required to undertake a study using this exposure assessment method. However, this approach is crude and does not account for the magnitude of emissions, incinerator characteristics, or the propagation of the emissions due to local meteorological and topographic conditions. Individual level direct measures of exposure, such as biomarkers in human tissue, provide an objective assessment of exposure to chemicals and are considered "gold standards" in exposure assessment [30]. Biomarkers are often not feasible in large studies due to the high cost of laboratory analysis, the difficulties in acquiring human tissue, and the burden and potential risks to participants involved [30]. Exposure modelling has largely bridged the gap between the need for more accurate exposure assessment and the practical and financial constraints of large epidemiological studies. Atmospheric dispersion models use monitored emission data along with information on local topographic and meteorological conditions, within a Gaussian framework, to estimate the concentration and dispersion pattern of pollutants around an identified source [31, 32]. New generation dispersion models have an updated understanding of atmospheric turbulence and boundary layer structure [33] and have been extensively evaluated [34–37].

Many studies investigating the relationship between incineration and adverse health outcomes have used distance as a proxy for exposure. Some studies have included additional information alongside proximity to strengthen this method, including wind patterns, soil concentrations [18], local topography, and complaints of nuisance caused by the plumes [24]. Only a limited number of more recent studies have used dispersion models [8, 12, 13, 17, 23] to assess exposures. As far as the authors are aware, no existing studies on incinerators have compared these two exposure assessment methods and quantified the extent of exposure misclassification between the two. Modelled exposure patterns are expected to be different when using the two comparative methods. The distance method will predict greatest exposure adjacent to the stack and will decrease linearly with distance from the stack. These exposures will also be fixed in time and will be homogenous in space at a given distance from the stack. In contrast, because stack height above ground is considered, the dispersion model will predict low concentrations of incinerator emissions near to the stack. Greatest concentrations will be at a distance from the stack (determined by the release conditions and meteorology) after which concentrations will decrease nonlinearly with distance. Temporal changes in release conditions and meteorology are taken into account to produce a concentration field that varies in time. Here, we provide a detailed comparison of atmospheric dispersion modelling and a distance based method to assess exposure to particulates from two MSWIs and explore issues of exposure misclassification.

4.2 METHODS

4.2.1 STUDY AREA AND STUDY POPULATION

Two UK MSWIs were included in this study, Crymlyn Burrows, located approximately 5 km east of Swansea, Wales and Marchwood, approximately 3 km west of Southampton, England. These two MSWIs are representative of operational MSWIs in Wales and England in terms of the operational standards they were built to (both have only ever operated to the most recent European Waste Incineration Directive [2]); their size (Crymlyn Burrows and Marchwood licensed throughput of 52,500 tonnes and 210,000 tonnes of MSW a year, respectively, where the typical median throughput of all operational UK MSWIs is 165,000 tonnes, ranging from 3,500 to 750,000 tonnes); and their rural locations (within 10 km surrounding Crymlyn Burrows 70% of the land is rural land and 69% for Marchwood, median for all operational MSWIs of 69%). The two selected incinerators additionally provided a number of contrasting features. Crymlyn Burrows has a single flue, is surrounded by hills, and lies 850 m from the coast, whereas Marchwood has two flues, is surrounded by flat land, and lies more inland. Incinerator characteristics and daily emissions data from their commissioning date (January 2003 for Crymlyn Burrows, January 2006 for Marchwood) until December 2010 were provided by the UK Environment Agency (EA).

The study area was defined as a 10 km radius around each MSWI. The 10 km distance was chosen for consistency with screening criteria used for implementing the Habitats Regulations: incineration plants that are within 10 km of a European Site require an assessment of their impact for short range air emissions.

The study population was defined as all residents within the study area, calculated by extracting postcode headcount data from the 2001 census [38], where one UK postcode represents on average 12–15 properties and 40–45 people.

4.2.2 EMISSIONS DISPERSION MODELLING

The Atmospheric Dispersion Modelling System Urban (ADMS-Urban) v2.3 modelling package was used [39] to model the dispersion pattern and ground level concentration of particles with a diameter <10 μm (PM_{10}) from both incinerators. ADMS-Urban is a new generation Gaussian plume air dispersion model that uses an updated understanding of turbulence and atmospheric boundary layer structure [33] and is capable of simulating the atmospheric dispersion patterns of pollutants from multiple sources and within complex terrain [40].

ADMS-Urban calculates atmospheric boundary layer parameters such as boundary layer height and Monin-Obukhov length from a variety of input parameters [40]: air temperature (°C), wind speed (m/s), wind direction (°), and cloud cover (oktas). The Monin-Obukhov length is an indicator of the atmospheric stability and is a key parameter in the dispersion of pollutant. It is defined by a quotient of heat flux at ground level by frictional velocity. It provides a height at which turbulent flows are created by buoyancy and not wind shear. In ADMS-Urban a minimum value for the Monin-Obkhov length is set, with the default value set to 30 m in order to account for the heat island effect of major cities and to prevent the model from stabilising [40, 41].

Another key model parameter that has impact on the dispersion of pollutants is the surface roughness length. Surface roughness length characterises the roughness of the terrain, providing an indicator of how much drag the wind experiences from the ground. Surface roughness is required to calculate convective turbulence.

TABLE 1: Source characteristics of the two inclusive municipal solid waste incinerators.

Incinerator	County	Permitted throughput (tonnes/ year)	Flue	Stack height (m)	Stack diam- eter (m)	Flue exit flow rate (m³/s)	Flue exit velocity (m/s)	Temp. (°C)
Crymlyn Burrows	Neath Port Talbot (South Wales)	52,500	1	40	0.95	12.3	17.6	136
March- wood	Hampshire (England)	210,000	1	65	1.25	30.3	24.7	150
			2	65	1.25	30.9	25.2	148

Flue exit flow rate, velocity, and temperature for Crymlyn Burrows provided are a mean of biannual measurements for most years of operation, whereas for Marchwood these measures are single measures derived from the permit application.

4.2.2.1 MODEL INPUT DATA

For each MSWI, information on the location of the stack, year commissioned, total annual waste licensed to incinerate and stack characteristics was extracted from their environmental permit application to the EA. The precise location of the stacks was verified by checking the incinerator address and postcode against six-figure grid references (georeferenced location of the stack in British National Grid projection), in addition to visually searching for stacks on satellite maps in Google maps. Stack data included number of lines, stack height (m), stack diameter (m), exit velocity (m/s), exit flow (m³/s), and exit temperature (°C) (Table 1). For Marchwood only one measure of flue gas flow, velocity, and temperature was available from 2006 till 2010. For Crymlyn Burrows quarterly measures of these flue gas metrics were available for most years of operation. Annual averages of these quarterly measures were calculated and used. When quarterly measures were unavailable, the overall representative flue gas measures for Crymlyn Burrows were used. The concentration of total particulates at the flue exit for each MSWI was measured as daily means.

Sensitivity analysis of the dispersion conditions was conducted to select the most appropriate and representative surface roughness and Monin-Obukhov lengths. The fetch for roughness is defined by the US Environmental Protection Agency (US EPA) as 1 km surrounding the source [42]. Land cover data, extracted from the CORINE Land Cover Map 2000 [43] (Figure 1), was used to characterise the 1 km area around each MSWI. CORINE is an EU-wide dataset, generated by semiautomatic classification of satellite imagery [43] and comprises 44 land cover classes, of which 11 relate to urban land. Based on the land cover data around each MSWI, an array of relevant lengths was selected. As both MSWIs were partly surrounded by urban land cover (Marchwood 20% and Crymlyn Burrows 26%, resp., see Figure 1), a number of different surface roughness lengths and minimum Monin-Obukhov lengths were explored. Output concentrations were then compared when using the different values for both lengths.

The surface elevation in the area surrounding the MSWIs was extracted from Ordnance Survey PANORAMA digital terrain model (DTM), which has a horizontal resolution of 50 m [44]. As shown in Figure 1 the terrain surrounding Marchwood is low lying with a mean elevation of 23 m above sea level. However for Crymlyn Burrows there is a significant variation in elevation, with a range of 370 m. In order to account for this variation in terrain and therefore changes in the dispersion pattern of particulates, the hill option in ADMS-Urban was selected and a preprepared terrain file was extracted from the DTM and input into the model.

Meteorological conditions greatly influence the observed spatial pattern of emitted pollutants from a point source. Selecting an appropriate meteorological station, that best represents the area surrounding the MSWI, is therefore crucial. Hourly land surface meteorological observations from all Met Office stations in England and Wales between 2003 and 2010 were obtained from the British Atmospheric Data Centre (BADC). Candidate meteorological stations located within approximately 30 km from the selected MSWIs were identified. Meteorological stations considered were those with 90% completeness for all weather variables (excluding cloud cover), for each year. The Air Quality Modelling Assessment

Unit (AQMAU) at the EA advised that incorporating cloud cover from alternative nearby stations makes a very small contribution to overall modelling uncertainties. Therefore, cloud cover was obtained from the nearest station with 90% completeness where necessary [45]. Following the selection of candidate meteorological stations, wind roses were plotted for each station. These wind roses were used to spot anomalies in the data (e.g., apparent gaps in wind from a given sector) and comparisons were made between the sites. Following this, CORINE land cover and DTM data were extracted and compared for a 1 km radius around each meteorological station in order to select a meteorological station with similar surrounding topography and land use to the MSWI. The dispersion model was then run using these different meteorological stations and their outputs compared.

4.2.2.2 MODEL OUTPUT

Bag-filtered stack emissions from the MSWIs were not considered to contain a significant amount of particulates greater than 10 μm diameter. Emitted particulates were therefore modelled as PM_{10} and considered to disperse in the same manner as a gas.

Modelled ground level concentrations of PM_{10} for the sensitivity analysis were estimated for receptors in a 200 m × 200 m grid within the study areas. For Marchwood sensitivity analysis was performed for 2006 and Crymlyn Burrows for 2003.

For the exposure analysis, all residential postcode centroids within the study area were used as receptors and ground level concentrations of PM_{10} were modelled. For Marchwood models were run for 2006–2010 and Crymlyn Burrows 2003, 2005–2010.

For the exposure analysis, each modelled day required input of single daily mean particulate concentrations at the flue exit together with hourly meteorological data to produce a daily ground level PM_{10} concentration field. These daily modelled concentrations were aggregated to calculate annual means. Model outputs were mapped in ESRI Arc-Map 10.0 [46].

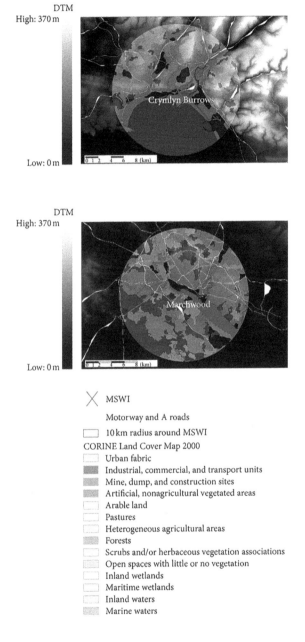

FIGURE 1: Land cover data from CORINE-Land 2001 and topography data from PANORMA 10 km around Crymlyn Burrows (a) and Marchwood (b) incinerators.

4.2.3 DISTANCE TO SOURCE

All residential postcode centroids within the study area were assigned a distance to their respective MSWI using the NEAR function in ArcGIS. The distance metric was chosen as distance from the edge of the study area rather than distance from the incinerator. This was termed proximity and had its greatest value at the incinerator and least value at the edge of the study area. The ordering of the magnitude of the proximity metric allowed a clearer comparison of the distance and dispersion approaches with the greatest proximity value and highest concentration found closest to the incinerator.

4.2.4 COMPARISON OF EXPOSURE ASSESSMENT METHODS

All residential postcodes within the study areas were assigned both an average modelled PM_{10} concentration over the period in which the MSWI was in operation and a distance to the MSWI. Postcodes were classified into deciles, quintiles, and tertiles from high to low exposures (modelled PM_{10} concentrations sorted from high to low, distance to MSWI from low to high). A population was additionally assigned to each postcode using headcount data extracted from the 2001 census [38].

The comparison of exposure assessment by the dispersion model and by the distance method was undertaken in three ways.

1. Calculation of Cohen's kappa coefficients of agreement between exposure deciles, quintiles, and tertiles as calculated by the distance method versus the dispersion model. Cohen's kappa coefficient provides a statistical measure of interobserver agreement taking into account chance, that is, a quantification of precision [47, 48]. Kappa coefficients range from 0 to 1, with 0 indicating no agreement and 1 perfect agreement between methods. As our exposure tertiles, quintiles, and deciles are ordinal categories, equal weighted kappa coefficients were calculated in addition to unweighted Cohen's kappa coefficients [49]. Weighted Cohen's kappa coefficients account for ordinal differences in categories;

that is, a difference of two categories between the indices of exposure is a more severe misclassification error than a difference of one category.

2. Calculation of weighted and unweighted Cohen's kappa coefficients of agreement between the distance method and the modelled particulate concentrations by population weighted exposure deciles, quintiles, and tertiles.

3. Plotting of modelled long-term average PM_{10} concentrations against distance from the MSWI at each postcode centroid, with calculation of Spearman's correlation coefficients.

4.3 RESULTS

4.3.1 PARTICULATE EMISSIONS FROM MSWI

Figures 2(a) and 2(b) display the daily concentrations of total particulates measured at the flue exit for Crymlyn Burrows and Marchwood, respectively. Figure 2(a) demonstrates the variability in concentrations for Crymlyn Burrows over the study period, 2003–2010, with the maximum concentration of $9.87 \, mg/m^3$. The gap in the data shown for 2004 was due to a fire during the last quarter of 2003 causing Crymlyn Burrows to stop operation during 2004. Figure 2(b) shows the daily particulate concentrations for both flues for the Marchwood incinerator. Again, there was considerable variability in concentrations over time and also between the two flues. Both Flue 1 and 2 had a maximum concentration of $10 \, mg/m^3$, the Waste Incineration Directive limit. Both MSWIs show a decreasing trend in particulate emissions from 2008 (Crymlyn Burrows) and 2009 (Marchwood) until 2010, from daily emissions of $\sim 10 \, mg/m^3$ to $1\text{-}2 \, mg/m^3$. The maximum particulate emissions took place in 2008 for both MSWIs.

4.3.2 DISPERSION MODELLING

For Marchwood, three candidate meteorological stations were located within 30 km. The nearest meteorological station was Southhampton

Oceanography Centre located 3.3 km east of Marchwood, followed by Solent (19.1 km south-east) and Middle Wallop (29.2 km north) (see Figure 3). For Crymlyn Burrows only one meteorological station was available located 9.4 km south-west from the incinerator.

Comparisons were made between the three meteorological stations available for Marchwood. First, the wind roses for the three meteorological stations were compared. The wind rose for the Southhampton Oceanography Centre displayed very low frequency of wind from the north-east, between 50 and 80 degrees, for all years of operation (2006–2010) (Figure 3(d)). The other two meteorological stations, however, did not show this pattern (Figures 3(b) and 3(c)). The effect of this apparent gap in wind direction becomes particularly evident when using the data from these meteorological stations in our dispersion model simulations. Figure 3 shows the modelled annual mean particulate concentrations in 2006 using the three meteorological stations around Marchwood MSWI. The PM_{10} annual mean concentration using the Southampton Oceanographic Centre clearly shows a gap in the predicted concentrations south-west of the incinerator (Figure 3(d)), not seen when using the other two meteorological stations (Figures 3(b) and 3(c)). Based on this comparison the data from Southampton Oceanographic Centre meteorological station was deemed erroneous for unknown reasons and was therefore not used in subsequent analysis. The wind and dispersion patterns were similar for Solent and Middle Wallop, with higher PM_{10} concentrations in the SW-NE diagonal. Therefore the closest station, Solent, was selected for the exposure analysis. However when Solent cloud cover fell short of 90% capture annually, cloud cover data from Middle Wallop was used.

An exploration of surface roughness for both MSWIs showed little variation in the model output for surface roughness lengths varying from 0.2 m to 1 m (see Figures 4(a) and 4(b)). Only 7.7% of the model receptors had a difference in modelled particulate concentrations greater than 25% in Marchwood and 3.1% for Crymlyn Burrows (Table 2). The difference in modelled particulate patterns and concentrations between no set minimum Monin-Obukhov length and 30m showed little variation, with a maximum percentage difference of 31% for Marchwood and 18% for Crymlyn Burrows (Table 2).

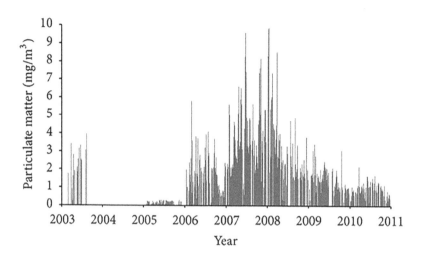

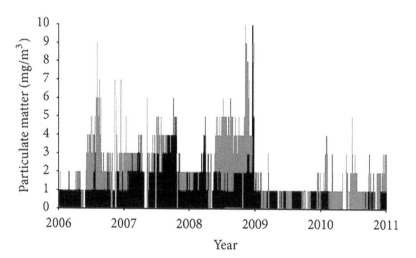

FIGURE 2: (a) Daily particulate concentrations measured at flue exit for Crymlyn Burrows from 2003 to 2010. (b) Daily particulate concentrations measured at flue exit for Marchwood from 2006 to 2010.

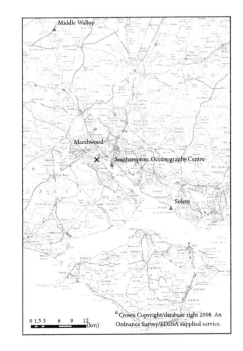

A

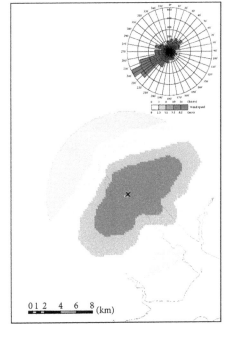

B

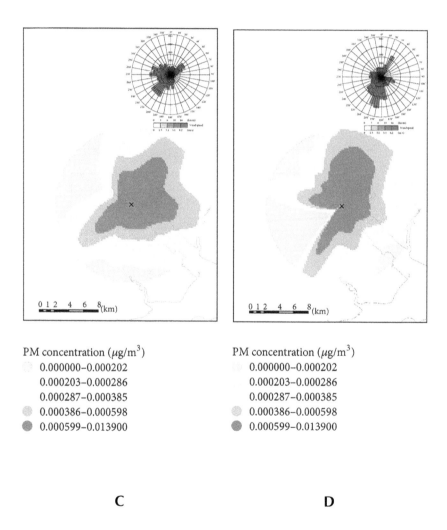

PM concentration (μg/m^3)
0.000000–0.000202
0.000203–0.000286
0.000287–0.000385
0.000386–0.000598
0.000599–0.013900

PM concentration (μg/m^3)
0.000000–0.000202
0.000203–0.000286
0.000287–0.000385
0.000386–0.000598
0.000599–0.013900

C   D

FIGURE 3: Sensitivity of the model to the selected meteorological stations for Marchwood in 2006. Maps (b)–(d) use the same site surface roughness length and minimum Monin-Obukhov length.

FIGURE 4 (A): Sensitivity of the model to site surface roughness length and minimum Monin-Obukhov length for Crymlyn Burrows. Maps (A)–(D) use the same meteorological station data for 2003.

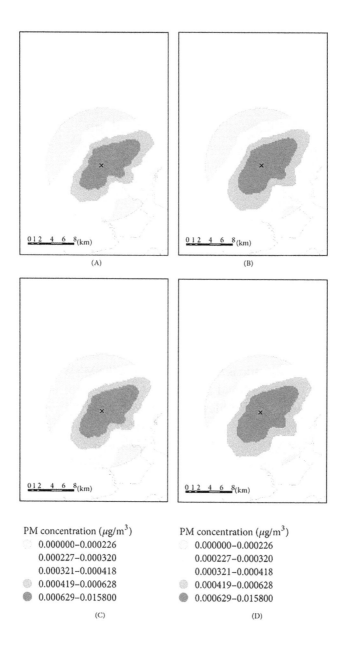

FIGURE 4 (B): Sensitivity of the model to site surface roughness length and minimum Monin-Obukhov length for Marchwood. Maps (A)–(D) use the same meteorological station data for 2006.

TABLE 2: Surface roughness sensitivity analysis. Percentage difference between extreme surface roughness values at all model receptors.

Percentage difference	Crymlyn Burrows	Marchwood
Surface roughness		
Mean (%)	8.7	12.3
Median (%)	6.9	11.2
Minimum (%)	0	0
Maximum (%)	116.6	117.5
Receptors > 25% difference (%)	3.1	7.7
Monin-Obukhov length		
Mean (%)	6.4	11.5
Median (%)	5.5	10.2
Minimum (%)	0	0
Maximum (%)	17.6	30.6
Receptors > 25% difference (%)	0	6.6

Table 3 demonstrates the extremely low concentrations of modelled annual PM_{10} concentrations within 10 km from the MSWI both for all days of the year (Table 3(a)) and also for only the days of operation (Table 3(b)). The modelled ground level concentrations of PM_{10} were extremely low for both MSWIs, with a mean concentration of $0.000117 \, \mu g/m^3$ for Crymlyn Burrows for all days and $0.000334 \, \mu g/m^3$ for operational days only; and $0.00129 \, \mu g/m^3$ for Marchwood for all days and $0.00205 \, \mu g/m^3$ for operational days only. Modelled long-term average PM_{10} concentrations were very small (maximum of $0.0022 \, \mu g/m^3$ for Crymlyn Burrows and $0.0089 \, \mu g/m^3$ for Marchwood). Figure 5 shows the modelled long-term average PM_{10} concentrations for both MSWIs against distance from the MSWI at each postcode centroid. It is clear from Figure 5 that the concentrations at the 10 km edge of the modelled domain were <7% of the maximum concentration.

The pattern of the final dispersion model for Crymlyn Burrows showed irregular shapes, with symmetrical bands of increasing exposure from the source. This irregular dispersion pattern might be due to the hilly topography in the Swansea area that modifies the wind patterns and, therefore, the dispersion of emissions from the MSWI. Due to its coastal location a large proportion of the modelled area has no population. For Marchwood the final dispersion pattern was much more elliptical with the greatest PM10 concentration extending to the north-east of the MSWI following the main wind direction.

4.3.3 DISTANCE TO SOURCE

Table 4 shows the number of postcodes and the population count in relation to distance from the two MSWIs. The area around the Marchwood MSWI is more densely populated (361,005 people within 10 km) than Crymlyn Burrows (248,937 people within 10 km). The population around Marchwood MSWI also resides closer to the MSWI than that at Crymlyn Burrows with the greatest population density between 3 km and 7 km.

TABLE 3 (A): Annual mean, median, and interquartile range of modelled PM_{10} concentration in the postcodes 10 km around Crymlyn Burrows (2003–2010) and Marchwood (2006–2010) weighted by postcode.

	Mean ($\times 10^5 \mu g/m^3$)	Median ($\times 10^5 \mu g/m^3$)	Interquartile range ($\times 10^5 \mu g/m^3$)
Crymlyn Burrows			
2003	3.7	3.0	2.5
2005	0.8	0.6	0.5
2006	10.8	8.2	6.8
2007	29.4	24.2	17.1
2008	22.0	16.5	15.2
2009	10.3	7.7	7.1
2010	4.9	3.9	3.0
Marchwood			
2006	121.5	80.0	101.6
2007	186.3	127.7	149.9
2008	229.5	139.5	200.8
2009	59.6	44.5	50.3
2010	48.9	36.4	32.2

TABLE 3 (B): Annual mean, median, and interquartile range of modelled PM_{10} concentration in the postcodes 10 km around Crymlyn Burrows (2003–2010) and Marchwood (2006–2010) weighted by postcode for operational days only.

	Mean ($\times 10^5$ $\mu g/m^3$)	Median ($\times 10^5$ $\mu g/m^3$)	Interquartile range ($\times 10^5$ $\mu g/m^3$)	Days of operation		
Crymlyn Burrows						
2003	51.5	4.25	31.7	33		
2005	2.61	2.10	1.59	150		
2006	29.1	23.6	17.1	204		
2007	61.2	51.1	33.6	264		
2008	52.7	42.6	32.6	227		
2009	23.1	18.8	13.8	225		
2010	13.3	10.6	7.77	188		
				Flue 1	Flue 2	Either one or both flues in operation
Marchwood						
2006	207.3	140.9	154.3	308	240	334
2007	290.6	204.1	221.5	325	327	344
2008	338.1	212.7	262.7	340	325	358
2009	96.9	72.5	72.0	296	192	357
2010	91.7	69.8	59.1	323	104	356

4.3.4 COMPARISON OF EXPOSURE ASSESSMENT METHODS

The agreement between exposure categories, as calculated by the dispersion modelling and distance methods, is shown in Table 5. Better agreement was achieved when using tertiles (Cohen's kappa coefficient of 0.424 unweighted and 0.553 weighted and 0.308 unweighted and 0.448 weighted from Crymlyn Burrows and Marchwood, resp.) compared with deciles and quintiles (Cohen's kappa coefficient ranging from 0.068 to 0.201 unweighted and 0.198 to 0.519 weighted).

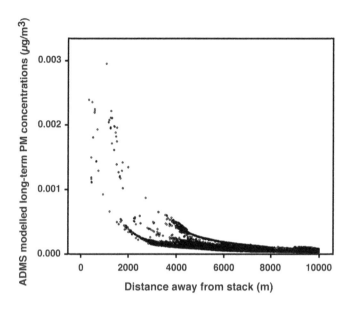

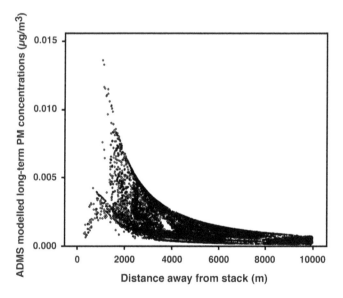

FIGURE 5: (a) Modelled long-term PM concentrations (μg/m³) plotted against distance away from the MSWI (m) at postcode centroid for Crymlyn Burrows. (b) Modelled long-term PM concentrations (μg/m³) plotted against distance away from the MSWI (m) at postcode centroid for Marchwood.

TABLE 4: Distance of the study population (all residents within 10 km) to the incinerators, Crymlyn Burrows and Marchwood.

Distance to source (km)	Crymlyn Burrow				Marchwood			
	Number of PCs	Percentage of total PCs (%)	Population count	Percentage of total population (%)	Number of PCs	Percentage of total PCs (%)	Population count	Percentage of total population (%)
0–<1	22	0.2	165	0.1	87	0.5	1677	0.5
1–<2	69	0.5	834	0.3	813	4.2	12829	3.6
2–<3	229	1.8	5067	2.0	2363	12.3	31729	8.8
3–<4	777	5.9	14590	5.9	2720	14.2	46690	12.9
4–<5	2623	20.1	38736	15.6	2969	15.5	59070	16.4
5–<6	2496	19.1	49338	19.8	2999	15.6	69784	19.3
6–<7	2365	18.1	51665	20.8	2171	11.3	48832	13.5
7–<8	1982	15.2	39467	15.9	1378	7.2	25298	7.0
8–<9	1256	9.6	25853	10.4	1611	8.4	32654	9.0
9–10	1251	9.6	23222	9.3	2055	10.7	32442	9.0
Total	13070	100	248937	100	19166	100	361005	100

PC: postcodes.

TABLE 5: Measure of agreement Kappa coefficient (where 0 = no agreement; 1 = perfect agreement) between modelled long-term PM_{10} concentrations and distance away from stack categorised in deciles, quintiles, and tertiles at postcode level.

	N	Type of Kappa	Deciles	Quintiles	Tertiles
Crymlyn Burrows	13069	Unweighted	0.0684	0.210	0.424
		Weighted-Equal	0.307	0.519	0.553
Marchwood	19166	Unweighted	0.0734	0.177	0.308
		Weighted-Equal	0.198	0.446	0.448

Table 6 shows the population weighted agreement of the two exposure methods. Again, agreement improved with a reduction in the numbers of exposure categories. Best agreement between methods was displayed for Crymlyn Burrows exposure tertiles (but here the unweighted Cohen's kappa coefficient only reached 0.425, equally weighted Cohen's kappa coefficient only reached 0.548) and the poorest agreement for Marchwood exposure deciles (unweighted Cohen's kappa coefficient 0.0644, equally weighted Cohen's kappa coefficient 0.150).

TABLE 6: Measure of agreement Kappa coefficient (where 0 = no agreement; 1 = perfect agreement) between population weighted modelled long-term PM_{10} concentrations and distance from the stack categorised in deciles, quintiles, and tertiles at postcode level.

	N	Type of Kappa	Deciles	Quintiles	Tertiles
Crymlyn Burrows	5269	Unweighted	0.0932	0.251	0.425
		Weighted-Equal	0.334	0.535	0.548
Marchwood	8102	Unweighted	0.0644	0.169	0.219
		Weighted-Equal	0.150	0.380	0.345

Figure 5 shows the long-term mean PM_{10} concentration at each postcode centroid against distance from MSWI for Crymlyn Burrows (Figure 5(a)) and Marchwood (Figure 5(b)). Spearman correlations (R) for modelled long-term PM_{10} concentrations versus proximity from the edge of the modelling domain at postcode level were 0.765 and 0.688 for

Crymlyn Burrows and Marchwood, respectively (both significant at the 0.01 level).

4.4 DISCUSSION

The majority of the studies exploring the relationship between incineration and health have used a simple distance metric as a proxy for exposure. Here we have provided a comparison of distance from source and emissions modelling to assess exposure to particulates emitted by two MSWIs in the UK. Our results suggest that epidemiological studies requiring an assessment of exposure to airborne pollutants from MSWIs, at a small scale level, would benefit from a dispersion modelling approach compared to a simple distance based approach. Although the use of distance as a proxy for exposure has limited data requirements, it does not account for source characteristics, the concentrations of pollutants emitted, local meteorological conditions, and topography [31, 50] all of which are incorporated in Gaussian dispersion models, such as ADMS-Urban. Dispersion models provide a different exposure assessment to distance from source. This approach is expected to be more realistic than a simple distance proxy as it tries to capture the physical processes that determine the dispersion of emissions from a point source including topographic and meteorological information that influence where and how emissions are dispersed. ADMS-Urban has been successfully used and validated when assigning exposure at an individual or small area level [34–37] and is frequently used for regulatory purposes, policy support, and providing information to the public [32]. Dispersion modelling can additionally help determine the distance to which a particular source influences exposures, as shown in Figure 5, where modelled PM_{10} falls to <7% of peak concentrations at 1000 m to 2000 m away from the MSWI. The comparison between dispersion modelling and distance for the two MSWIs studied here (see Table 5 and Figure 5) reveals poor to moderate agreement only when using distance compared with dispersion modelling. Both methods assigned a decreasing exposure with an increasing distance from source (as shown by the strong spearman's correlations with continuous measures). However, when using categorical metrics (as are often employed in epidemiological studies) distance was a fairly good proxy in distinguishing highest and lowest exposure tertiles, but the dispersion model was able to capture the pattern of small area level variation in population exposure (Figures 3 and 5), which did not conform to circular dispersion around the source as would be predicted using a distance model.

The influence of stack height on the dispersion pattern was especially apparent for the Marchwood MSWI, which shows very small PM_{10} concentrations up to approximately 500 m (Figure 5(b)), after which they peak between 1000 and 2000 m, depending on the direction. This pattern was less apparent at the Crymlyn Burrows MSWI, mainly due to the lack of postcodes within 2000 m of the MSWI.

Both Figures 5(a) and 5(b) show a flattening in modelled PM_{10} concentrations beyond approximately 5 km, suggesting that, at least for the Crymlyn Burrows and Marchwood MSWI, most variability in exposure occurs within 5 km of a MSWI and this was therefore captured well within the 10 km distance chosen in this assessment.

Model input parameters influenced both the pattern and concentration of the modelled PM_{10}, in turn affecting the modelled exposed population. It is therefore essential that the quality of the model input parameters is assessed. It was found that the model was sensitive to surface roughness length, Monin-Obukhov length, and meteorological conditions. The model output showed little relative variation in output concentrations with different input parameters with the exception of changes in meteorological station. We demonstrate here that the choice of meteorological input data is crucial. As shown in Figure 3, possible misclassification of exposure is evident from the use of different meteorological stations, particularly in the case of the south-west part of the Marchwood MSWI.

The dispersion model simulations in this study were subject to a number of limitations that would contribute to the uncertainty in the ground level exposure estimates produced. Firstly, Marchwood only had a single measure of flue gas flow, velocity, and temperature for the duration of its operation (2006 till 2010), whereas Crymlyn Burrows had quarterly measures of these flue gas metrics for most years of operation which showed substantial variation. The assumption that these flue exit parameters are constant over such long periods of time is therefore not representative of true conditions. Additionally, due to data availability, poor data quality, and completeness, the choice of meteorological sites was limited and it was challenging to find meteorological sites representative of the surrounding area. This was especially evident for Marchwood where the selected meteorological site (Solent) was located 19 km away from the MSWI. Additionally, although ADMS-Urban has been validated as a point source modelling tool in other scenarios, the long-term mean concentrations of modelled PM_{10} in this study were exceptionally low, and therefore model validation would not be possible, as they fall below the limit of detection for regulatory ambient measurements.

There are a number of disadvantages to using dispersion models, including their large input data demands, which are often unavailable, and the expertise required to successfully run and interpret the models [31]. To meet the EU Directive requirements the MSWIs in this study, along with those elsewhere in Europe, are now required to have daily measurements of particulate emissions. This allows time varying emissions to be included in modelled assessments for the first time. This is beneficial for calculating exposures linked to health endpoints with critical exposure periods, for example, trimester specific exposures for birth outcomes.

Although long-term ground level PM_{10} levels from these MSWIs were found to be approximately thousandths of regional background levels, it is hypothesised that particulates from MSWIs may possibly have different impacts on health than those from other ambient sources of particulate matter due to their metal or dioxin content, for instance. The modelled concentrations of PM10 may act as a proxy for the concentration fields for these and other primary emissions from MSWIs. While long-term PM_{10} concentrations from dispersion modelling may provide a good indication of ambient concentrations, this will still be an imperfect marker of personal exposure. An alternative individual level exposure could be measured by personal monitoring or collection and analysis of biomarkers. However, such personal exposure approaches, aside from being very expensive and time-consuming and (for biomarkers) potentially invasive, may not adequately capture exposures specific to MSWIs.

4.5 CONCLUSIONS

Using distance as a proxy measure of exposure to emissions from incinerators is a simple, quick, and cheap approach; however, when compared with dispersion modelling, there is indication of exposure misclassification. Dispersion models incorporate information on individual incinerator characteristics, emission concentrations, local meteorological conditions, and topography, all of which contribute to the observed concentrations and spatial patterns of incinerator emissions. The additional detail included in these models enables a more appropriate and informative exposure assessment from incinerators, which is important in an epidemiological context in order to reduce risk of bias in risk estimates due to exposure misclassification.

REFERENCES

1. European Union, "Landfill Directive, 1999/31/EC," Official Journal of the European Communities, 1999.
2. The European Parliment and The council of the European Union, "Waste Incineration Directive (2000/76/EC)," Official Journal of the European Communities, 2000.
3. Department for Environment Food and Rural Affairs (DEFRA), "Waste hierarchy," https://www.gov.uk/waste-legislation-and-regulations.
4. D. Crowley, A. Staines, C. Collins et al., "Health and environmental effects of landfilling and incineration of waste—a literature review," Tech. Rep. Paper 3, 2003.
5. L. Rushton, "Health hazards and waste management," British Medical Bulletin, vol. 68, pp. 183–197, 2003.
6. Committee on Carcinogenity of Chemicals in Food Consumer Products and the Environment, "Cancer incidence near municipal solid waste incinerators in Great Britain," 2000.
7. Committee on Carcinogenity ofChemicals in Food Consumer Products and the Environment, "Update statement on the review of cancer incidence near municipal solid waste incinerators," 2009.
8. S. Cordier, A. Lehébel, E. Amar et al., "Maternal residence near municipal waste incinerators and the risk of urinary tract birth defects," Occupational and Environmental Medicine, vol. 67, no. 7, pp. 493–499, 2010.
9. M. Franchini, M. Rial, E. Buiatti, and F. Bianchi, "Health effects of exposure to waste incinerator emissions: a review of epidemiological studies," Annali dell'Istituto Superiore di Sanita, vol. 40, no. 1, pp. 101–115, 2004.

10. D. Porta, S. Milani, A. I. Lazzarino, C. A. Perucci, and F. Forastiere, "Systematic review of epidemiological studies on health effects associated with management of solid waste," Environmental Health, vol. 8, no. 1, article 60, 2009.

11. J.-F. Viel, N. Floret, E. Deconinck, J.-F. Focant, E. De Pauw, and J.-Y. Cahn, "Increased risk of non-Hodgkin lymphoma and serum organochlorine concentrations among neighbors of a municipal solid waste incinerator," Environment International, vol. 37, no. 2, pp. 449–453, 2011.

12. M. Vinceti, C. Malagoli, S. Fabbi et al., "Risk of congenital anomalies around a municipal solid waste incinerator: a GIS-based case-control study," International Journal of Health Geographics, vol. 8, no. 1, article 8, 2009.

13. S. Cordier, C. Chevrier, E. Robert-Gnansia, C. Lorente, P. Brula, and M. Hours, "Risk of congenital anomalies in the vicinity of municipal solid waste incinerators," Occupational and Environmental Medicine, vol. 61, no. 1, pp. 8–15, 2004.

14. P. A. Cresswell, J. E. S. Scott, S. Pattenden, and M. Vrijheid, "Risk of congenital anomalies near the Byker waste combustion plant," Journal of Public Health Medicine, vol. 25, no. 3, pp. 237–242, 2003.

15. T. J. B. Dummer, H. O. Dickinson, and L. Parker, "Adverse pregnancy outcomes around incinerators and crematoriums in Cumbria, north west England, 1956–93," Journal of Epidemiology and Community Health, vol. 57, no. 6, pp. 456–461, 2003.

16. B. Jansson and L. Voog, "Dioxin from Swedish municipal incinerators and the occurence of cleft lip and palate malformations," International Journal of Environmental Studies, vol. 34, pp. 99–104, 1998.

17. C.-M. Lin, C.-Y. Li, and I.-F. Mao, "Birth outcomes of infants born in areas with elevated ambient exposure to incinerator generated PCDD/Fs," Environment International, vol. 32, no. 5, pp. 624–629, 2006.

18. O. L. Lloyd, M. M. Lloyd, F. L. R. Williams, and A. Lawson, "Twinning in human populations and in cattle exposed to air pollution from incinerators," British Journal of Industrial Medicine, vol. 45, no. 8, pp. 556–560, 1988.

19. N. Obi-Osius, B. Misselwitz, W. Karmaus, and J. Witten, "Twin frequency and industrial pollution in different regions of Hesse, Germany," Occupational and Environmental Medicine, vol. 61, no. 6, pp. 482–487, 2004.

20. H. Rydhstroem, "No obvious spatial clustering of twin births in sweden between 1973 and 1990," Environmental Research, vol. 76, no. 1, pp. 27–31, 1998.

21. T. Tango, T. Fujita, T. Tanihata et al., "Risk of adverse reproductive outcomes associated with proximity to municipal solid waste incinerators with high dioxin emission levels in Japan," Journal of Epidemiology, vol. 14, no. 3, pp. 83–93, 2004.

22. G. W. t. Tusscher, G. A. Stam, and J. G. Koppe, "Open chemical combustions resulting in a local increased incidence of orofacial clefts," Chemosphere, vol. 40, no. 9-11, pp. 1263–1270, 2000.

23. M. Vinceti, C. Malagoli, S. Teggi et al., "Adverse pregnancy outcomes in a population exposed to the emissions of a municipal waste incinerator," Science of the Total Environment, vol. 407, no. 1, pp. 116–121, 2008.

24. F. L. R. Williams, A. B. Lawson, and O. L. Lloyd, "Low sex ratios of births in areas at risk from air pollution from incinerators, as shown by geographical analysis and

3-dimensional mapping," International Journal of Epidemiology, vol. 21, no. 2, pp. 311–319, 1992.

25. Health Protection Agency, "The impact on health of emissions to air from municipal waste incinerators," in Position Statement on Municipal Solid Waste Incinerators, 2009.

26. P. Elliott and D. Wartenberg, "Spatial epidemiology: current approaches and future challenges," Environmental Health Perspectives, vol. 112, no. 9, pp. 998–1006, 2004.

27. K. Steenland and D. Savitz, Topics in Environmental Epidemiology, Oxford University Press, Oxford, UK, 1997.

28. B. G. Armstrong, "Effect of measurement error on epidemiological studies of environmental and occupational exposures," Occupational and Environmental Medicine, vol. 55, no. 10, pp. 651–656, 1998.

29. D. Baker and M. J. Nieuwenhuijsen, Environmental Epidemiology: Study Methods and Applications, Oxford University Press, Oxford, UK, 2008.

30. K. Sexton, L. L. Needham, and J. L. Pirkle, "Human biomonitoring of environmental chemicals: measuring chemicals in human tissues is the "gold standard" for assessing exposure to pollution," American Scientist, vol. 92, pp. 38–45, 2004.

31. M. Jerrett, A. Arain, P. Kanaroglou et al., "A review and evaluation of intraurban air pollution exposure models," Journal of Exposure Analysis and Environmental Epidemiology, vol. 15, no. 2, pp. 185–204, 2005.

32. N. Moussiopoulos, E. Berge, T. Bohler et al., "Ambient air quality, pollutant dispersion and transport models," Topic Report no. 19/1996, European Environment Agency, 1996.

33. D. Briggs, "Exposure assessmentin," in Spatial Epidemiology: Methods and Applications, P. Elliott, J. Wakefield, N. Nest, and D. Briggs, Eds., Oxford University Press, Oxford, UK, 2000.

34. D. J. Carruthers, H. A. Edmunds, A. E. Lester, C. A. McHugh, and R. J. Singles, "Use and validation of ADMS-Urban in contrasting urban and industrial locations," International Journal of Environment and Pollution, vol. 14, no. 1-6, pp. 364–374, 2000.

35. S. R. Hanna, B. A. Egan, J. Purdum, and J. Wagler, "Evaluation of the ADMS, AERMOD, and ISC3 dispersion models with the OPTEX, Duke Forest, Kincaid, Indianapolis and Lovett field datasets," International Journal of Environment and Pollution, vol. 16, no. 1-6, pp. 301–314, 2001.

36. B. Owen, H. A. Edmunds, D. J. Carruthers, and D. W. Raper, "Use of a new generation urban scale dispersion model to estimate the concentration of oxides of nitrogen and sulphur dioxide in a large urban area," Science of the Total Environment, vol. 235, no. 1–3, pp. 277–291, 1999.

37. S. Righi, P. Lucialli, and E. Pollini, "Statistical and diagnostic evaluation of the ADMS-Urban model compared with an urban air quality monitoring network," Atmospheric Environment, vol. 43, no. 25, pp. 3850–3857, 2009.

38. Office for National Statistics, "Census 2001: postcode headcounts," 2004.

39. Cambridge Environmental Research Consultants, ADMS-Urban, Cambridge, UK, 2008.

40. Cambridge Environmental Research Consultants, "ADMS-Urban: an urban air quality management system," User Guide, Cambridge, UK, 2010.
41. D. J. Carruthers, R. J. Holroyd, J. C. R. Hunt et al., "UK-ADMS: a new approach to modelling dispersion in the earth's atmospheric boundary layer," Journal of Wind Engineering and Industrial Aerodynamics, vol. 52, no. C, pp. 139–153, 1994.
42. AEROMOD Implementation Workgroup. US Environmnetal Protection Agency, AEROMOD Implementation Guide, 2009.
43. European Environment Agency, "CORINE Land Cover (CLC2000)," 2005, http://www.eea.europa.eu/publications/COR0-landcover.
44. Ordnance Survey, "Land-Form PANORAMA," http://www.ordnancesurvey.co.uk/oswebsite/products/land-form-panorama/index.html.
45. N. Bettinson, "Air quality modelling assessment unit I.C. London," London, 2012.
46. ESRI, "ArcMap," Redlands, Calif, USA, 2010.
47. J. Cohen, "A coefficient of agreement for nominal scales," Educational and Psychological Measurement, vol. 20, no. 1, pp. 37–46, 1960.
48. A. J. Viera and J. M. Garrett, "Understanding interobserver agreement: the kappa statistic," Family Medicine, vol. 37, no. 5, pp. 360–363, 2005.
49. J. Cohen, "Weighted kappa: nominal scale agreement provision for scaled disagreement or partial credit," Psychological Bulletin, vol. 70, no. 4, pp. 213–220, 1968.
50. T. Bellander, N. Berglind, P. Gustavsson et al., "Using geographic information systems to assess individual historical exposure to air pollution from traffic and house heating in stockholm," Environmental Health Perspectives, vol. 109, no. 6, pp. 633–639, 2001.

PART III

INDUSTRIAL EMISSIONS

CHAPTER 5

Comparative Analysis of Monitoring Devices for Particulate Content in Exhaust Gases

BEATRICE CASTELLANI, ELENA MORINI, MIRKO FILIPPONI, ANDREA NICOLINI, MASSIMO PALOMBO, FRANCO COTANA, AND FEDERICO ROSSI

5.1 INTRODUCTION

Over the past century, scientists and environmental regulators have focused on particulate matter (PM) as one of the major areas of air pollution study and control.

Particulate matter is released as particles and includes ash, dust or rapidly agglomerating aerosols from various industrial processes via stack emissions to air [1]. The main sources of particulate include the combustion of coal, oil, gasoline/petrol, diesel, wood, biomass and high temperature industrial processes, such as smelters and steel mills.

The composition of particulate matter is highly variable and may include substances such as sulfates, nitrates, hydrogen ions, ammonium, elemental carbon, silica, alumina, organic compounds, trace elements, trace metals, particle bound water and biogenic organic species [2].

The subject of particulate continuous emission monitoring to satisfy regulatory requirements is of relatively new interest as a result of recent changes in legislation. With the advent of emission limits defined in terms of mass concentration (expressed in mg/m^3), instead of in terms of color or opacity as in the past, the issue of continuous particulate monitoring has become a new and growing regulatory requirement [3].

Operators of industrial stacks use continuous particulate monitoring instrumentation for a variety of process and environmental purposes: (i) to provide better feedback on a process, (ii) to provide continuous control, (iii) to satisfy environmental legislation. Therefore, particulate emission monitoring can be categorized by the quality and type of information provided.

Gross failure detection or broken bag detection is the simplest form of particulate monitoring since it is just a qualitative monitoring. An alarm is activated to detect a significant increase of particulate loading, indicating a filter failure. Instruments used for filter failure detection do not necessarily need to be accurate, nor have the sensitivity to measure dust levels in normal conditions. In these cases, there is no regulatory need to calibrate the instrument since the output is in terms of a relative dust output rather than an absolute level. Units of measurement are usually a percentage of full scale or a factor of normal emissions [4].

For concentration measurements in mg/m^3 aimed at assessing the compliance with the relevant directives, the absolute level of particulate is the issue of critical importance and the instrument must provide a calibrated output on a continuous basis. Calibration gives in situ continuous emission monitors the ability to monitor particulate in absolute terms. It consists of isokinetic or gravimetric sampling in which a sample of flue gas is collected and weighed.

Particulate emission monitoring is a challenging technical field, not only because of the specific accuracy and performance of particulate monitors, but also due to the harsh environment in which they must continuously operate.

Several studies available in literature focus on in-field tests of commercially available particulate matter continuous emission monitoring systems (PM CEMS) in industrial applications. Since the adoption of a technology should be driven by its effectiveness and value in the targeted application, the aim of the present paper is to provide a comparative analysis of the currently available technologies for measuring particulate releases to atmosphere. To do this, after a description of the relevant legislation, an overview of the main industrial stationary sources and a description of the main types of sampling systems offered by manufacturers are presented. The techniques most commonly used for particulate monitoring are opacity, dynamic opacity, light scattering, beta attenuation, triboelectric and electrodynamic.

5.2 RELEVANT LEGISLATION

In this paragraph, the relevant European Directives and Legislation as well as technical standards are reported.

The three core EU Directives affecting industrial processes are:

- the Large Combustion Plant Directive (LCPD) 2001/80/EC which regulates large power plant, oil refineries and boiler plant on large industrial complexes;
- the Waste Incineration Directive (WID) 2000/76/EC, with relevance to incineration processes and processes that use waste as a fuel source; this covers cement kilns, some metallurgical processes and some renewable energy plant.
- the IPPC or Integrated Pollution Prevention and Control, which provides a framework for regulating chemicals, metals and minerals processes as well as defined permits for combustion and incineration plant.

The proposed Industrial Emissions Directive would consolidate these distinct requirements in a single piece of European legislation. Emission limits and monitoring requirements are defined directly in the Directives or derived from the framework defined in the industry specific BREF (Best Available Technique Reference) documents and country specific interpretation of BREF notes which are written to support the IPPC directive.

In Europe, processes falling under the Waste Incineration Directive (WID) and Large Combustion Plant Directive (LCPD) must continuously monitor particulate emissions in mg/m^3 in compliance with the European Standard EN 14181. The instruments are calibrated by comparison to a reference isokinetic sampling method in compliance with EN 13284 part 1 or ISO 9096.

Relevant technical standards for particulate emissions monitoring are shown in Table 1.

TABLE 1: Technical standards for particulate emissions monitoring.

Standard	Content
DIN EN 13284-1	Stationary source emissions—Determination of low range mass concentration of dust—Part 1: Manual gravimetric method
DIN EN 13284-2	Stationary source emissions—Determination of low range mass concentration of dust—Part 2: Automated measuring systems
DIN EN 14181	Stationary source emissions—Quality assurance of automated measuring systems
ISO 23210	Stationary source emissions—Determination of PM10/PM2,5 mass concentration in flue gas—Measurement at low concentrations by use of impactors
ISO 9096	Stationary source emissions—Manual determination of mass concentration of particulate matter

5.3 OVERVIEW OF STATIONARY EMISSION SOURCES

Major stationary sources, such as electric power plants, oil processing plants, cement production plants, and municipal waste combustors, pose several environmental problems, also in terms of dust emission. In addition, continuous monitoring systems at these sources often work in arduous operating conditions. This fact makes it important to understand the nature and concentration of the emitted dust.

Therefore, this paragraph provides an overview of the main stationary industrial emission sources and a description of emitted particulate matter features. In terms of particulate emissions, one of the most problematic processes in the petroleum refining chain is the fluid catalytic cracking (FCC) of heavier fractions, because of their higher hetero-atom concentration, metal contents and coking tendency. Fluid catalytic cracking (FCC) is used in the oil refining industry to convert heavy fractions to lighter products. Several process and catalyst innovations have been made to tackle the above-mentioned problems. A new generation of FCC catalyst technology has emerged with tailor-made catalysts for higher structural stability and attrition strength, more complete CO combustion during re-generation, reducing SOx emissions from FCC stacks [5,6].

Loss of catalyst is a major source of dust emission in FCC sections. The catalyst used in the FCC process is produced in the form of fine pow-der usually below 180 μm. It comprises of 5%–40% zeolite in a matrix of alumina, semisynthetic clay derived gel or natural clay [7]. In their study of the microstructure of FCC, Bass and Ritter [8] described in great detail the chemical composition and morphology of recently developed cata-lysts, which are a combination of gel, clay and zeolite.

Loss of catalyst powder has been receiving attention for highly abra-sive dust emissions [9,10,11,12]. The highly abrasive dust produced in FCC is critical for the installation of proper continuous emission moni-toring systems. Data on particulate monitoring systems in FCC units are given by Antwerp Total Refinery [13].

At the Antwerp Total Refinery, both FCC units are operating in partial combustion mode. Part of the coke remains on the catalyst and therefore it is burned and partially converted into CO_2 and CO. CO-rich gas passes via cyclones to a downstream boiler where the combustion is completed, generating high pressure steam.

Before 2005, catalyst particles passed the CO boiler unchanged and were emitted via stack to the environment. Since 2005, an electrostatic precipitator (ESP) has been removing the majority of these particles.

For the installation of the ESPs, the refinery installed a PM CEMS for ESP monitoring and legal compliance. The installed PM CEMS was provided by Sick/Maihak (Germany), type FW56-I-Ex. The measurement principle is based on light absorption.

To avoid dust abrasion and deposits, PM CEMS was supplied with flushing air to keep the optical parts free. Since this aspect was proven critical, the system was modified. Currently, air supply to the mirror is independent from that to the transmitter/receiver and each one is equipped with its own flow indicator [13].

The use of advanced duct monitoring technologies is a high priority also for operators of waste incinerators and much data on the application of PM CEMS for compliance with the particulate emission standard of waste combustors can be found in the literature.

Eli Lilly and Company conducted a demonstration of commercially available PM CEMS on a liquid hazardous waste incinerator at Lilly's Clinton Labs in Clinton, Indiana. The objective of this demonstration was to evaluate the performance and reliability of PM CEMS in a moisture-saturated flue gas over several months of operation [14].

The company had the primary objective of determining how to make instrumentation work accurately in their applications. Technical concerns were primarily related to application in a wet flue gas [15]. The two instruments used in this test were the Sigrist (model KTNRM/SIGAR 4000) and the Groupe Environment S. A. (ESA) Model Beta 5M.

Results showed that the selected PM CEMS required significant, unit-specific operation and development time in order to achieve acceptable calibration. The initial failure of the ESA unit to operate properly supported the need for an initial break-in period. The endurance data for the ESA and Sigrist units were encouraging. Uptime of the ESA and Sigrist monitors were near or above the suggested requirements [15].

Another field study to evaluate the performance of three commercially available PM CEMS was conducted at the US Department of Energy (DOE) Toxic Substances Control Act (TSCA) Incinerator [16]. The three monitors were Durag F-904 K beta monitor, The Environment SA Beta 5M (ESA) and Sigrist CTNR extractive light-scattering monitor.

Several important conclusions were drawn from the results of this field study. The light scattering device required only minor maintenance and operated trouble-free throughout the study, while the beta gauge monitors had several operational problems and required a more rigorous maintenance. The beta gauge that reported emissions on a dry basis was particularly hampered with problems arising from condensation formation. Results from this test however establish the suitability of beta gauge technology for monitoring PM emissions from incinerators [16].

Other types of stationary emission sources are stacks attached to the raw mill, rotary kiln, coal mill, grate cooler, cement mill in a cement plant [17].

Majority of particulates emitted from cement industry may range from 0.05 to 5.0 μm in diameter [18]. In both wet and dry process plants with dust control technology, about 85% of escaping particles were less than 10 μm in diameter, while in dry plants having bag houses, about 45% of escaping particles was of 2.5 μm diameter [19,20]. The particulate matter contains elemental content (Ca^{2+}, NO_3^-, SO_4^{2-}, As, Cd, Co, Cr, Cu, Fe, Mn, Ni, Pb and Zn) of the principal raw materials, products, combustion material from the kiln stack in a cement plant [21]. Among the elements of environmental concern (As, Cd, Cr, Ni, Pb), As, Cd and Pb showed higher concentration in stack emitted particles [21]. Another noteworthy characteristic of the aerosol from cement plants is that its size distribution is very stable [20].

In the steel industry, PM CEMS have been used for providing qualitative information on the operation and maintenance of filter bags, but not so much for the quantitative estimation of emissions [22]. These plants are considered major sources of PM_{10} and $PM_{2.5}$ emissions. A significant reduction of particulate emissions of sinter plants, until 5 mg/Nm³, can be achieved by fabric filters on a continuing basis [23].

There are studies in the literature for testing the applicability of continuous PM CEMS for quantitative evaluation of steel plants' emissions. In [22] different continuous emission monitoring systems for PM were compared in field conditions at a steel melting shop. The tests were performed using four commercially available monitoring instruments based on probe electrification and light scattering. Results of the tests showed that the compared instruments were not suitable for the quantitative estimation of dust emissions in widely varying field conditions. Another problem concerning the use of these monitors in quantitative measurement of emissions was the calibration of continuous PM concentration monitors. PM concentrations below 2 mg/m³, which predominate in the steel melting shop for most of the time, cannot be measured very reliably [22].

Coal combustion has been recognized as one of the major sources of fine particulates. Morphological analysis shows that the PM from pulverized coal-fired plant is composed of regular, spherical particles. In contrast, PM from circulating fluidized bed plants consists of particles of various shapes, including agglomerates of spherical, flake-like and floccus-like particles [24].

TABLE 2: Data on emissions of major stationary sources.

Process	PM size and concentration	PM composition	PM CEMS	References
FCC—refinery	~180 μm	5%–40% zeolite in a matrix of alumina (highly abrasive)	light absorption (critical aspect related to dust abrasion and deposits on optics)	[7,8,13]
Cement plants	0.05–5.0 μm	Elemental content (Ca^{2+}, NO_3^-, SO_4^{2-}, As, Cd, Co, Cr, Cu, Fe, Mn, Ni, Pb, Zn)	probe electrification and light scattering (problems with calibration for concentration below 2mg/m³)	[18,21,22]
		Elements of environmental concern (As, Cd, Cr, Ni, Pb)		
Coal combustion	PM 50 (54.7%)	pulverized coal-fired plant: regular, spherical particles	-	[24,25]
	PM 10 (19.9%)			
	PM 2.5 (1.3%)	circulating fluidized bed plants: agglomerates of spherical, flake-like and floccus-like particles		
Biomass combustion	0–100 kW: 14.4 mg/Nm³	-	-	[26]
	100–350 kW: 34.8 mg/Nm³			
	350 kW–1 MW: 57.5 mg/Nm³			
	1 MW–2 MW: 67.0 mg/Nm³			
	350–2.000 kW: 61.2 mg/Nm³			
	2–5 MW: 9.4 mg/Nm³			
	> 5 MW: 10.9 mg/Nm³			

Particulate emissions from coal-fired power stations with high efficiency ESPs result in concentration lower than 100 mg/m^3. The size distribution shows that PM 50 constitute 54.7% of total dust, while PM 10 and PM 2.5 respectively 19.9% and 1.3% [25].

As far as biomass combustion plants are concerned, it has been shown that small combustion boilers for district heating have considerably lower emission values than limits in regulations [26,27].

Cyclones and ESP (in larger installations) are used as abatement technologies. It is interesting that emissions are highest for the medium sized boilers [26]. While smaller boilers (<2 MW) use multi-cyclones only, larger boilers (>2 MW) have to apply ESP to meet the emission limit value (50 mg/Nm3 for boilers >2 MW). Low emission values of the smallest boilers are most likely because only wood chips are used as fuels, compared to saw dust and wood wastes in medium sized boilers.

Concerning the split of the size category 350 kW–2 MW it can be seen that the average values for boilers in the category 350 kW to 1 MW are approximately 10 mg/Nm^3 lower than those of the category 1 to 2 MW [26]. Data on PM emissions of the analyzed stationary sources are summarized

5.4 PARTICULATE MATTER CONTINUOUS EMISSION MONITORING TECHNOLOGIES

The main analytical principles used in instruments to measure dust concentrations are described below. These principles are opacity, light scattering, beta attenuation, probe electrification (triboelectric effect, electrodynamic device).

PM CEMS based on such technologies must be calibrated by gravimetric and isokinetic sampling to provide a continuous output of dust concentration in mg/m^3. In fact, gravimetric sampling is the only method that gives real concentration. Gravimetric measurement consists in taking off a partial gas flow via a filter head probe. The dust content is determined by weighting the dust collector mass before and after extraction. Gravimetric sampling is carried out isokinetically: it means that the collected particles have the same velocity in the sampling nozzle as elsewhere in the stream. This increases the accuracy and reliability of results.

FIGURE 1: Opacity measurement setup [29].

5.4.1 OPACITY

Opacity meters measure the decrease in light intensity due to absorption and scattering as the beam crosses the stack according to Beers-Lambert's Law. The basic operational principle of these instruments is that a collimated beam of visible light is directed through a gas stream toward receiving optics (Figure 1). The receiving optics measure the decrease in light intensity, and the instrument electronics convert the signal to an instrument output. Technical description of commercial opacity meters is given in Table 3. These instruments measure smoke density in transmission, opacity, Ringelmann units or optical density (extinction) and/or mass concentration of particulate in mg/Nm^3 [28].

The intensity of the light at the detector, I, is compared with the reference light intensity, I_o, to give the transmittance T, as shown in Equation (1):

$$T = I/I_o \tag{1}$$

Transmittance can be converted to opacity Op (Equation (2)) or optical density D (Equation (3)):

$$Op = 1 - T \tag{2}$$

$$D = \log (1/T) \tag{3}$$

The loss of light intensity can be correlated to particulate mass concentration measured by manual gravimetric sampling.

There are two formats for opacity devices. Single path monitors simply project a beam across a duct to a receiver. Dual beam devices have a reflector mirror on the opposite side of the stack from the light source and the beam is projected between two transceivers. This enables each transceiver to compensate for gradual window contamination by using clean mirrors inserted periodically into the beam path. In this way, any errors caused by misalignment of the sensors may be compensated for.

The dual-pass opacity meter allows all the instrument electronics to be incorporated into one unit. Incorporating the light source and detector into one instrument also allows direct measurement of the loss of light. In fact, the source intensity and the loss of light are measured and compared at the same time. This helps prevent inaccurate readings due to the degradation of the light source intensity that is a common problem in basic meters.

A opacity meter used as PM CEMS should use a red or near infrared light source, and not the white light source used on traditional opacity monitors since the extinction-to-mass concentration for a given aerosol type is dependent on particle size within the visible light spectrum but nearly independent of particle size at the infrared wavelength. Some manufacturers have started using a green LED to monitor both opacity and PM concentration simultaneously [30].

Opacity measurements are dependent on particle size, composition, shape, color and refractive index. These properties may change with fuel type and thus calibration is necessary with variation of process conditions [31,32]. In general, the measurement sensitivity of opacity meters is not fine enough to detect small changes in PM concentration.

An alternative type of cross stack optical dust monitor is the dynamic opacity device. While traditional opacity instruments measure the intensity of received light, the dynamic opacity technology instead measures the ratio of signal scintillation to absolute light intensity. This offers a significant advantage over traditional opacity methods, as the ratiometric measurement is unaffected by lens contamination allowing the instrument to operate with lens contamination exceeding 90%. In fact, since both the reduction in light intensity and the variation in intensity caused by lens contamination are affected by the same proportion, it results in no net effect. This therefore greatly reduces costly process intervention for lens maintenance and servicing.

The dynamic opacity device is suitable for stacks after bag filters, cartridge filters, cyclones, electrostatic precipitators, variable flue gas velocities, including low velocity flue gases, variable particulate size and type [29].

5.4.2 LIGHT SCATTERING

Scattering is due to reflection and refraction of the light by the particle. The amount of light scattered is based on the concentration of particles and the properties of the particles in the light's path (e.g., size, shape, and color of the particles) [37,38]. If the wavelength of the incident light is much larger than the radius of the particle, a type of scattering called Rayleigh scattering occurs. If the wavelength of the incident light is about the same size as the radius of the particle, Mie scattering will occur (Figure 2).

A light scatter instrument measures the amount of light scattered in a particular direction (forward, side, or backward) and outputs a signal proportional to the amount of particulate matter in the stream. The dust concentration is derived by correlating the output of the instrument to manual gravimetric measurements [4].

Some components included in these instruments to minimize the effect of interference and degradation of the light source are: (i) the use of a pulsed light and (ii) parallel measurement of the light source intensity. The use of the pulsed light source limits the possibility of other sources' interference, because the instrument only measures the reflected light while the instrument light source is on. The parallel measurement of the light source intensity accounts for degradation of the light source because a reference of the source intensity is measured along with each scattered light measurement.

TABLE 3: Technical data of commercial opacity meters [33,34,35,36].

Model	STACK 602	DR 290	DR 220	LAND 4200	LAND 4500 III	DUSTHUNTER T200
Measurement	Ratiometric opacity technology	Optical transmission of visible light	Optical transmission of visible light	Path transmissometry	Path transmissometry	Transmittance measurement
Sensors	Cross-stack	Double pass	Double pass	Double pass	Cross stack, double pass	Cross-duct
Light Source	Modulated LED (green spectrum)	Wide Band Diode-White SWBD LED white, 450–680 nm	LED, green 530 nm	High intensity LED red 623 ± 20 nm	High Intensity LED Green 520 ± 20 nm	Not available
Duct/stack diameter	1–15 m	1–18 m	0.4–15 m	0.3–9.7 m	0.5–10 m	0.5–12 m
Max Temperature flue gas	600 °C	600 °C	600 °C	600 °C	600 °C	600 °C
Measurement range	10–1,000 mg/m³	0.5–15/ 500–10,000 mg/m³	2–10,000 mg/m³	0–100/0–999 mg/m³	0–10/ 0–10,000 mg/m³	0–200/ 0–10,000 mg/m³
Standard Compliance	EN 14181 EN 13284-2	EN 13284-2 EN 14181 EN 15267-3	EN 14181	EN 15267-1 EN 15267-2 EN 15267-3 EN 14181	EN 14181 EN 15267-3	
Comments	applications with electrostatic precipitator large diameter emission stacks variable flue gas velocities, including low velocity flue gases variable particulate size and type	suitable for systems with variable gas speed super-wide band diode (SWBD) reduces influence of variable particle sizes	filter monitoring suitable for applications with variable gas speed	process/non-compliance performance reduced for pathlengths >7.5 m	measurement independent of gas velocity, humidity and particle charge	

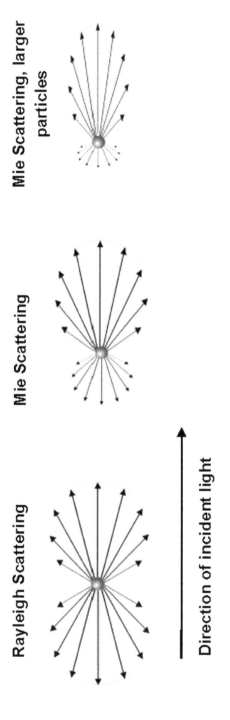

FIGURE 2: Mie and Rayleigh scattering.

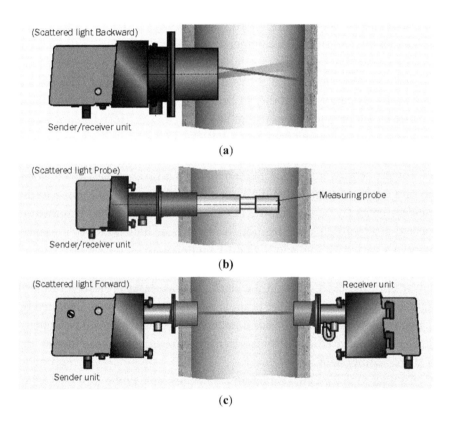

FIGURE 3: Light-scattering configurations: (a) backward scattering (b) probe forward scattering (c) cross forward scattering [36].

For scattered light measurements, back and forward scattering are used. They are shown in Figure 3. Back scatter devices are particularly suitable for in situ applications in small ducts, where low levels of dust are present. Low angle of back scatter measurement increases the effective penetration of the measurement volume into the stack but makes the instrument less sensitive to fine dust.

There are three types of forward scatter devices available currently: (i) the extractive type, (ii) probe configuration and (iii) cross duct configuration.

The extractive type draws a sample from the stack via a sampling nozzle and then presents it to a forward scattering photometer. The advantage of this system is the ability to heat the sampling system, where there are significant amounts of moisture in the stack. The sensor measures the amount of light scattered back from particles in the stack illuminated by a modulated laser [36].

The probe forward scatter instrument has a measurement volume at the tip of a probe and measures the light scattered at a forward angle to the incident beam (typically coming from a laser diode). If the instrument is located in a representative position, it can provide high accuracy measurement in a variety of low and high dust applications.

The cross duct forward scatter instrument has a transmitter and a receiver opposite each other on the stack. A diode laser projects a beam of light into the stack: part of the beam is attenuated and some is scattered by the particulate. The receiver has a large lens behind which are two photodetectors, the nearer lens detects a transmission signal and the further, the scattered component.

Compared to forward scatter cross stack designs, probe forward scatter provides a representative measurement without the errors deriving from misalignment, vibration, near wall measurement sensitivity and the complexity of keeping a double head system clean. In addition, cross stack scatter has a varying response to dust along its measurement path [4].

A technical description of commercial light scattering meters is given in Table 4.

TABLE 4: Technical data of commercial light scattering meters [33,34,36,39].

Model	STACK 181	DR 300-40	DR 800	SB50	SB100	C200	SF100	SP100	FEW 200	SIGRIST STACK-GUARD
Principle	Low-angle forward light scattering	Backward light scattering (halogen lamp white)	Forward light scattering (laser diode, red 650nm)	Scattered light backward	Scattered light backward laser wavelength between 640 nm and 660 nm forward	Combination of light forward transmission Cross-duct version	Scattered light forward probe version (extractive type for wet gases)	Scattered light forward (extractive type)	Scattered light forward light (extractive type)	Scattered light wavelength 650 nm
Duct/stack diameter (multi-sensor configuration required for stack >3m)	250mm–3m	>0.3 m	>0.3 m	>0.5 m	<0.5 m	0.5–8 m	0.5–3 m 2.5–6 m	≥0.25 m	-	-
Max flue gas temperature	250 °C (optional 500 °C)	320 °C	220 °C	600 °C	600 °C	300 °C	300 °C	400 °C	220 °C	160 °C
Measurement range	0–15/0–100 mg/m³	0.5–10/10–200 mg/m³	0.5–10/10–200 mg/m³	0–20/0–200 mg/m³	0–10/0–200 mg/m³	scattered light: 0–5/0–200 mg/m³ transmission 0–200/0–10,000 mg/m3	0–5/0–200 mg/m³	0–5/0–200 mg/m³	0–5/0–200 mg/m³	0–100 mg/m³ PLA (polystyrene-latex-aerosol)

TABLE 4: *Cont.*

Model	STACK 181	DR 300-40	DR 800	SB50	SB100	C200	SF100	SP100	FEW 200	SIGRIST STACK-GUARD
Standard Compliance	EN 15267-3 EN 14181 EN ISO 14956	EN 13284-2 EN 14181	EN 13284-2 EN 14181	EN 15267-3 EN 14181	EN 15267-3 EN 14181	EN 15267-3 EN 14181 DIN ISO 14956	EN 14181 EN 15267	EN 14181 EN 15267	EN 14181 EN 15267	EN 14181
Comments	after electrostatic precipitator both constant and variable flue gas velocities variable particulate size	variable stack gas speed low to medium dust concentration	variable stack gas speed low to medium dust concentration	low to medium concentration	low to medium concentration	very low and high dust concentration	very low to medium dust concentration small to medium duct diameters	very low to medium dust concentration small to medium duct diameters	very low to medium dust concentration gas sampling and return combined in one probe	steam-saturated and corrosive gases

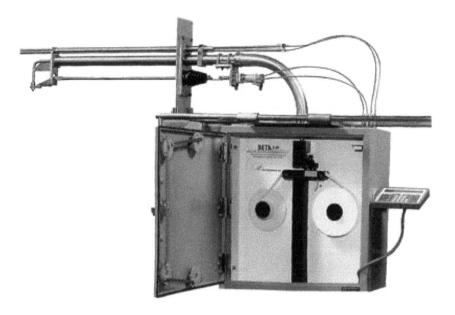

FIGURE 4: Beta attenuation for dust monitoring on stacks [40].

5.4.3 BETA ATTENUATION

β-gauge samplers are the only systems which continuously measure the mass concentration of particulate by extraction. The particles are collected isokinetically on a filter tape and the change in transmission of β-particles from a radioactive source is monitored. The particulate laden gas is extracted via a small nozzle from the duct. The extraction rate is controlled by a duct flow sensing system. The captured material is placed on a constantly moving sticky tape and then presented to a β gauge to measure the mass (Figure 4).

The two main components of a beta attenuation measuring system are the beta source, in general Carbon-14, and the detector. Many different types of detectors can quantify beta particle counts, but the ones most widely used are the Geiger Mueller counter or a photodiode detector.

Beta systems do not provide short term dynamic monitoring of particulates and a single point measurement may not always be representative. The heated isokinetic sampling train is prone to maintenance problems. Measurements are made against a reference measurement already on the tape in mg/m^3.

The advantage is that they are not affected by chemical composition, size or color changes in the particles, and the use of a heated probe obviates water effects. Technical description of commercial Beta attenuation meters is given in Table 5.

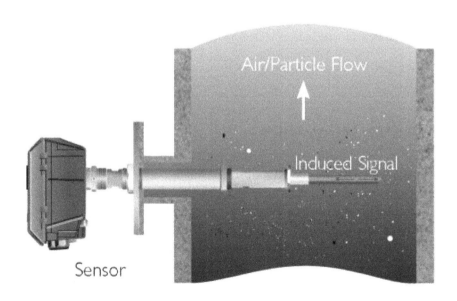

FIGURE 5: Probe Electrification device [33].

TABLE 5: Technical data of commercial beta attenuation meters [34,41].

Model	F-904-20	BETA 5M
Principle	Beta gauge measurement with Isokinetic sampling	Beta gauge measurement with Isokinetic sampling
Source		very low activity Carbon 14 source
Duct/stack diameter	>0.5 m	-
Velocity	-	4 to 40 m/s
Max flue gas temperature	0–250 °C, optional up to 500 °C	170 °C
Measurement range	0–1/0–1000 mg/m3	2–4000 mg/m^3
Standard Compliance	EN 14181	ISO 10473
		EN 13284-2
		EN 14181
		ISO 9096
Comments	unaffected by particle size, color or moisture	independent of particulate characteristics
	measuring of very low emission dust concentration	
	small diameter stack monitoring of dust concentration	

5.4.4 TRIBOELECTRIC EFFECT

Triboelectric devices (Figure 5) detect three separate effects when particulate strikes or passes close to a conductor placed in a particle laden gas stream: (i) when a particle strikes the conductor, a charge transfer takes place between particle and conductor; (ii) as the particle strikes the conductor it rubs on the surface and causes a frictional charge; (iii) as charged particles pass close to the conductor they induce a charge of equal and opposite magnitude in the conductor. The amount of charge generated by the first two effects depends on the velocity of the particle, its mass and the charge history of the particle, while the third effect is an inductive charge. The size of the charge is dependent on the proximity of the particle to the conductor and the charge history of the particle [40].

Since the response of the probe is sensitive to gas velocity, these systems are most suited to situations where the gas flow is fairly constant.

Probe electrification does not work well in wet gas streams with water droplets or when the particles are subject to a varying electrical charge.

Triboelectric monitors are very sensitive to low levels of particulate concentration. They work best where the particulate material is non-conductive. Like other dust monitors, this system has to be calibrated against an extractive method at each individual site.

TABLE 6: Technical data of commercial electrification devices [33,34,43].

Model	PFM 02 V	D-RX 250	QAL 991	VIEW 370
Principle	Measurement with triboelectric sensor	Dust: measuring the transfer of electrical charge from dust particles to an electrode in flowing sample gas	ElectroDynamic Probe Electrification technology	ElectroDynamic Probe Electrification technology
		Flow: measuring the differential pressure created by a multi point pitot tube		
Duct/stack diameter		>0.3 m	0.5–3 m (multi-sensor configuration required for stack >3m)	0.1–6 m
Flow velocity	from 3 m/s	7–35 m/s	8–20 m/s	8–20 m/s
Max flue gas temperature	280 °C	350 °C	250 °C/500 °C	800 °C
Measurement range	0–10/0–1000 mg/m³	0–10/0–500 mg/m³	0–1000 mg/m³	0–500 mg/m³
Comments			constant velocity required outside its velocity range	constant velocity required outside its velocity range

5.4.5 ELECTRODYNAMIC DEVICE

Like other probe electrification devices, the sensor measures the current created by particles passing and colliding with a grounded sensor rod inserted into stack.

The sensor electronics filter out the DC current created by particle collisions on the rod and measure an RMS signal within an optimized frequency bandwidth which results from the particles passing the rod. This signal, being independent of the rod surface condition, has a stable and repeatable relationship to dust concentration in many types of industrial applications.

Since the signal is not dependent on particle collisions (unlike triboelectric) the related problems of rod contamination and velocity dependence are minimized [42]. In applications where the particle charge, particle size and particle distribution remain constant the resulting alternating current is proportional to dust concentration.

Technical description of commercial electrification devices is given in Table 6.

5.5 COMPARISON AND CONCLUSIONS

The advent of emission limits, expressed in mg/m^3, requires the use on industrial stacks of PM CEMS. The comparative analysis presented in this paper is driven by the fact that there are a variety of industrial processes which produce dust emissions in the environment. To satisfy legislation and industrial requirements, a full range of techniques are used in practice and provide a practical and robust solution for most industrial applications.

The most used PM CEMS in industrial applications, in accordance with the analyzed literature papers, are light scattering devices, opacity meters and electrification devices. Extractive types such as beta gauge are less used than in situ types.

One of the fundamental issues in obtaining good results from particulate instruments is to ensure that the instrument is fit for purpose for the intended application [7,8,9,10,29]. As a result of the analysis carried out in the previous paragraph, Table 7 shows the core application areas of the different technologies.

Scattering instruments in general can measure much lower emissions than opacity instruments and are therefore suitable for processes controlled by highly efficient bagfilters [44]. Compared to backscatter, opacity and dynamic opacity systems, probe forward scattering technique may be used to accurately monitor very low dust concentrations.

As far as electrification devices are concerned, if compared to opacity systems, they do not suffer from misalignment and are suitable for measuring dust levels below 0.1 mg/m³. In case of particle charging by electrostatic precipitators, electrification technologies are outside their application limits and light-scattering can provide an alternative solution. A comparison of opacity, light scattering and electrification, based on stack diameter and PM concentration is given in Figure 6.

Concluding, the performance and suitability of any particulate monitor is application dependent [29]. Each type of CEMS presents disadvantages or advantages over other types of CEMS for a targeted industrial application. The choice of a PM CEMS for a plant should be driven by the correlation between operating parameters and proper technical characteristics of PM CEMS.

TABLE 7: Comparison of particulate monitoring technologies.

Measurement Technology		Stack diameter (m)	Concentration (mg/m³)		Filter Type	Velocity dependent
			Min	Max		
Probe Electrification	Triboelectric		0	1000	Bag, Cyclone, Drier, Scrubber (no water droplets), None	No (for 8–20 m/s)
	Electrodynamic	0.5–3	0	1000	Bag, Cyclone	Yes
Transmisometry	Ratiometric Opacity	1–15	10	1000	Bag (concentration dependent), Cyclone, EP, None	No
	Opacity	0.5–18	0	10000	EP, None	No
Scattered light	Forward	0.25–6	0.1	200	Bag, Cyclone, EP	No
	Back	0.3–4	0.5	200	Bag, Cyclone, EP	No

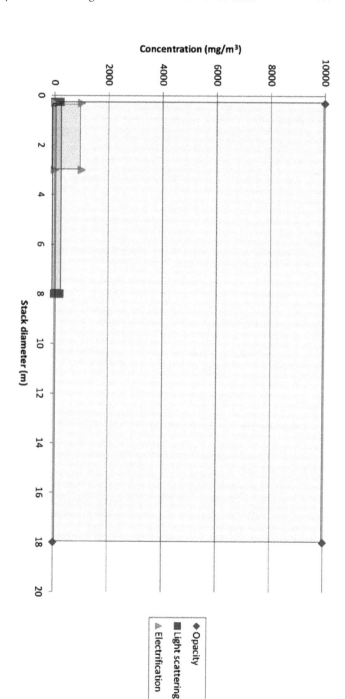

FIGURE 6: Comparison of opacity, light scattering and electrification: stack diameter and concentration.

REFERENCES

1. Engel-Cox, J.; Thi, N.; Oanh, K.; van Donkelaar, A.; Martin, R.V.; Zell, E. Toward the next generation of air quality monitoring: Particulate Matter. Atmos. Environ. 2013, 80, 584–590.
2. Environment Agency. Technical Guidance Note (Monitoring) M15-Monitoring PM 10 and PM 2.5; Environment Agency: Rotherham, UK, 2012; p. 16.
3. England, G.C.; Zielinska, B.; Loos, K.; Crane, I.; Ritter, K. Characterizing PM 2.5 emission profiles for stationary sources: Comparison of traditional and dilution sampling techniques. Fuel Process. Tech. 2000, 65, 177–188.
4. United States Environmental Protection Agency. Current Knowledge of Particulate Matter (Pm) Continuous Emission Monitoring; US Office of Air Quality: Research Triangle Park, NC, USA, 2000.
5. Otterstedta, J.E.; Geverta, S.B.; Jääsb, S.G.; Menona, P.G. Fluid catalytic cracking of heavy (residual) oil fractions: A review. Appl. Catal. 1986, 22, 159–179.
6. Boerefijna, R.; Guddeb, N.J.; Ghadiria, M. A review of attrition of fluid cracking catalyst particles. Adv. Powder Tech. 2000, 11, 145–174.
7. Kirk, R.E.; Othmer, D.F.; Grayson, M.; Eckroth, D. Kirk-Othmer Encyclopaedia of Chemical Technology, 3rd ed.; John Wiley: New York, NY, USA, 1981; p. 664.
8. Bass, J.L.; Ritter, R.E. The microstructure of fluidizable cracking catalysts. J. Mater. Sci. 1977, 12, 583–594.
9. Johnsson, F.; Breitholz, C.; Leckner, B. Solids segregation in a CFB-boiler furnace. In Fluidization IX; Fan, L.-S., Knowlton, T.M., Eds.; Engineering Foundation: New York, NY, USA, 1998; pp. 757–764.
10. Forsythe, W.L., Jr.; Hertwig, W.R. Attrition characteristics of fluid cracking catalysts—laboratory studies. Ind. Eng. Chem. 1949, 41, 1200–1206.
11. Wu, C.; Cheng, Y.; Ding, Y.; Jin, Y. CFD-DEM simulation of gas—solid reacting flows in fluid catalytic cracking (FCC) process. Chem. Eng. Sci. 2010, 65, 542–549.
12. Jiménez-García, G.; Aguilar-López, R.; Maya-Yescas, R. The fluidized-bed catalytic cracking unit building its future environment. Fuel 2011, 90, 3531–3541.
13. Schacht, J.; Courtheyn, J. ESP units realise major dust emission reduction at Total Refinery Antwerp. In Proceedings of Dust Conf 2007, International Conference, Maastricht, The Netherlands, 23–24 April 2007.
14. Bastian, R.E.; Lambert, R.H. Particulate matter continuous emission monitor test performance in a moisture-saturated flue gas. Waste Manag. 2000, 20, 369–377.
15. Keener, M.E.; Lambert, R.H.; Bastian, R.E. The effect of averaging time on compliance of a PM threshold when using continuous emission monitors. Waste Manag. 2000, 20, 379–383.
16. Dunn, J.E., Jr.; Davis, W.T.; Calcagno, J.A.; Allen, M.W. Field testing of particulate matter continuous emission monitors at the DOE Oak Ridge TSCA incinerator. Waste Manag. 2002, 22, 427–438.
17. Chen, C.; Habert, G.; Bouzidi, Y.; Jullien, A. Environmental impact of cement production: detail of the different processes and cement plant variability evaluation. J. Clean. Prod. 2010, 18, 478–485.

18. Kalacic, I. Chronic nonspecific lung disease in cement workers. Arch. Environ. Health 1973, 26, 78–83.
19. Van Oss, H.G.; Padovani, A.C. Cement Manufacture and the Environment Part II: Environmental Challenges and Opportunities. J. Ind. Ecol. 2003, 7, 93–126.
20. Fraboulet, I.; Chaucherie, X.; Gouriou, F.; Gautier, F.; Karoski, N.; Thille, H.; Fiani, E.; Le Bihan, O. Aerosol size distribution determination from stack emissions: the case of a cement plant. In Proceedings of Dust Conf 2007, International Conference; Maastricht, The Netherlands: 23–24 April 2007.
21. Gupta, R.K.; Majumdar, D.; Trivedi, J.V.; Bhanarkar, A.D. Particulate matter and elemental emissions from a cement kiln. Fuel Proc. Tech. 2012, 104, 343–351.
22. Kangas, R. Continuous monitoring of exhaust gas dust emissions from a steel melting shop. Filtrat. Separ. 2004, 41, 35–40.
23. Fleischandel, A.; Plattnel, T.; Lanzerstorfer, C. Efficient Reduction of PM 10/2.5 emissions at Iron Ore Sinter Plants. In Proceedings of Dust Conf 2007, International Conference, Maastricht, The Netherlands, 23–24 April 2007.
24. Yao, Q.; Li, S.-Q.; Xu, H.-W.; Zhuo, J.-K.; Song, Q. Studies on formation and control of combustion particulate matter in China: A review. Energy 2010, 35, 4480–4493.
25. Meij, R.; te Winkel, B.H.; Spoelstra, H.; Erbrink, J.J. Aerosol emissions from dutch coal-fired power stations. In Proceedings of Dust Conf 2007, International Conference, Maastricht, The Netherlands, 23–24 April 2007.
26. Krutzler, T.; Böhmer, S. Dust emissions from biomass boilers in Austria. In Proceedings of Dust Conf 2007, International Conference, Maastricht, The Netherlands, 23–24 April 2007.
27. Yoo, J.-I.; Kim, K.H.; Jang, H.N.; Seo, Y.; Seok, K.S.; Hong, J.H.; Jang, M. Emission characteristics of particulate matter and heavy metals from small incinerators and boilers. Atmos. Environ. 2002, 36, 5057–5066.
28. Horne, R. Particulate emissions—Optical and other methods for continuous monitoring from a point source. In Industrial Air Pollution Monitoring, 1st ed.; Springer: Houten, The Netherlands, 1998; pp. 61–80.
29. Averdieck, W. Selection of Particulate Monitors; Technical Paper 4 (Issue 1). PCME Ltd.: St Ives, Cambridgeshire, UK, 2011; p. 8.
30. Uthe, E.E. Evaluation of an Infrared Transmissometer for Monitoring Particulate Mass Concentrations of Emissions from Stationary Sources. J. Air Poll. Control Assoc. 1980, 30, 382–386.
31. Conner, W.D. Measurement of Opacity and Mass Concentration of Particulate Emissions by Transmissometry; EPA 650/2-74-128. National Environmental Research Center, Office of Research and Development, U.S. Environmental Protection Agency: Research Triangle Park, NC, USA, 1974.
32. Conner, W.D.; Knapp, K.T.; Nader, J.S. Applicability of Transmissometers to Opacity Measurement of Emissions—Oil-fired Power Plants and Portland Cement Plants; EPA 600/2-79-188. U.S. Environmental Protection Agency: Washington, DC, USA, 2002.
33. PCME. Available online: http://www.pcme.com/ (accessed on 28 May 2014).
34. Durag Group. Available online: http://www.durag.com/d_r_290.asp (accessed on 28 May 2014).

35. Land Ametek Process and Analytical Instruments. Available online: http://www.lan-dinst.com (accessed on 28 May 2014).

36. Sick Sensor Intelligence. Available online: http://www.sick.com/group/en/home/products/ (accessed on 28 May 2014).

37. McCartney, E.J. Optics of the Atmosphere: Scattering by Molecules and Particles; John Wiley and Sons, Inc.: New York, NY, USA, 1976; p. 421.

38. Redmond, H.E.; Dial, K.D.; Thompson, J.E. Light scattering and absorption by wind blown dust: Theory, measurement, and recent data. Aeolian Res. 2010, 2, 5–26.

39. Sigrist Process Photometer. Available online: http://www.photometer.com/en/index.html (accessed on 30 June 2014).

40. Matsusaka, S.; Maruyama, H.; Matsuyama, T.; Ghadiri, M. Triboelectric charging of powders: A review. J. Electrostatics 2004, 62, 277–290.

41. Environment SA. Available online: http://www.environnement-sa.com/products-page/ (accessed on 28 May 2014).

42. Averdieck, W. Electrodynamic Technology for Particulate Monitoring; Technical Paper 13 (Issue 12). PCME Ltd.: St Ives, Cambridgeshire, UK, 1999; Volume 5.

43. Dr. Födisch Umweltmesstechnik AG. Available online: http://www.foedisch.de/en/products/devices/dust-measurement-devices/r-dust-measurement-devices.html (accessed on 28 May 2014).

44. Peeler, J.W.; Jahnke, J.A. Handbook: Continuous Emission Monitoring Systems for Non-Criteria Pollutants; EPA 625/R-97/001. U.S. Environmental Protection Agency: Washington, DC, USA, 1997.

CHAPTER 6

Perspectives of Unconventional PCDD/F Monitoring for a Steel Making Plant

ELENA CRISTINA RADA, MARCO RAGAZZI, GABRIELA IONESCU, MARCO TUBINO, WERNER TIRLER, AND MAURIZIO TAVA

6.1 INTRODUCTION

In the last years, the scientific community has concentrated their research on dioxin source and their impact over surroundings. PCDD/Fs can be released from natural causes by incomplete combustion of organic material or volcanic activities. Evidence shows that human activities are the most main generator and polluter of dioxin with: steel sector [1, 2], cement works [3] and waste treatments [4, 5, 6]. In European Union the introduction of more and more stringent emission limits has significantly reduced the role of incineration in the emission inventories. Some studies have analysed the PCDD/Fs presence in soil [7], ambient air [8], food fats

*Perspectives of Unconventional PCDD/F Monitoring for a Steel Making Plant. Rada EC, Ragazzi M, Ionescu G, Tubino M, Tirler W, amd Tava M. Stiinta si Inginerie, **2014**,26, (2014). http://stiintasiinginerie.ro/wp-content/uploads/2014/07/26-1.pdf. Reprinted with permission from the authors.*

[9], human blood and breast milk [10, 11]. Ashes, sediments and sewage sludge have been analysed in many countries [12, 13] in order to have a clear overview of the PCDD/Fs presence.

These chemicals were, and remain, of interest because of the extremely high toxicity of the congener 2,3,7,8 tetrachlorodibenzopdioxin (2,3,7,8-TCDD), and the demonstrated or estimated high toxicity of 16 other dioxin congeners [14]. The legislative framework is quite controversial and according to WHO, the Tolerable Daily Intake (TDI) of PCDD/F and dl-PCB is 1-4 $pg_{TEQ}/kg_{body\ weight}$. The compliance of that value reduces the health effects [15] under a threshold considered statistically acceptable.

The aim of this study is to point out the role of PCDDFs monitoring in a case-study related to a steel making plant in a valley in the North of Italy where PCDD/Fs resulted the pollutants potentially more impacting on the territory [16, 17]. In particular, the role of unconventional monitoring through deposimeters and soil characterization is discussed in details.

6.2 MATERIAL AND METHODS

The atmospheric deposition of PCDD/Fs was measured with Depobulk® deposimeters [16]. This tool has its body in glass and is able to capture the overall deposition of PCDD/Fs in air in a planned period, generally no longer than one month. PCDD/Fs were analyzed in a laboratory in accordance with the EPA method 1613B

For the analytic determination of PCDD/Fs in soil, some samples were taken from the sites near the steal making plant. The analytic quantification was made after the sample purification, through high resolution gas chromatography—high resolution mass spectrometry (HRGC-HRMS), using the EPA method 1613B.

The selection of the monitoring sampling sites was performed using a modelling of the conveyed and diffused emissions of the steel making plant [18, 19]. The selected model was AERMOD, coupled with preprocessors aimed to the characterization of the local meteorology. As the deposimeters detect the present impact of the plant, whilst soils can point out criticalities related to the past impact, modelling was developed with data both concerned the present authorized plant configuration and the past one. In Figure 1 an example of the modelling results is shown. The black rectangle refers to the plant surface.

The availability of maps of impact on the territory allowed the choice of sites significant for distance, population presence, wind exposure. Three sites were selected for PCDD/F deposition characterisation: two are located eastbound and one westbound. Concerning soils, ten sites where selected at progressive distance from the plant in different directions.

A comprehensive view of the generated data is reported in [19]. In the present paper those data are shown in Tables 1 and 2, referring to the significance of the plant impact.

The area of interest is located in a narrow valley, where the dilution of emissions is significantly different in summer compared to winter. Moreover, winter domestic heating can contribute to additional depositions of PCDD/F in the area [19,20]. These phenomena can partially explain the monthly variability of data.

Concerning deposimeters, an increase of the deposition of PCDD/F from plant closed periods to plant operating periods was not always detected. From another point of view it must be noticed that the material used as feed can vary, thus contributing to the variability of the phenomena. The ratio between total PCDD/F and TEQ pointed out a high variability of the values. That can depend on the fact that in the area there is not a dominant source of PCDD/F. The data obtained in the site closest to the plant pointed out the importance of the diffused emissions. However, the average values in the two sites closer to the plant remained below the limits proposed at international level. In particular, a deposition threshold limit was developed for the specific area [14]. The present configuration of the plant, supported by the best available technologies of the sector and in case of good management, seems to be able to comply with that limit, assessed through the TDI.

Data in the farthest site resulted more complex to be analyzed. Some values were significantly higher than the ones normally reported, not dependently on the operation of the plant. The phenomenon can be explained by the presence of another significant PCDD/F source.

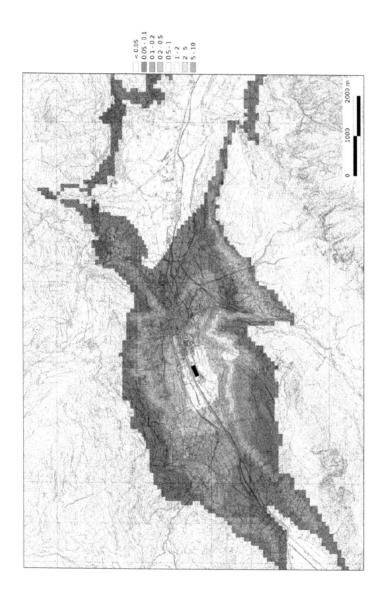

FIGURE 1: Example of conveyed and diffused (ng_{I-TEQ} m^{-2} y^{-1}) [19]

FIGURE 2: Selected sites for future monitoring

TABLE 1

PCDD/F deposition ($pg_{I\text{-}TEQ}$ m^{-2} y^{-1}) closest site	974
PCDD/F deposition ($pg_{I\text{-}TEQ}$ m^{-2} y^{-1}) intermediate site	491
PCDD/F deposition ($pg_{I\text{-}TEQ}$ m^{-2} y^{-1}) farthest site	15,583
Distance correlation from the plant	Partial
Average compliance with diet threshold limit [14]	2 sites out of 3

Concerning soil characterization, a comprehensive overview of the generated data is reported in [19]. In the present article those data are summarised in Table 2 referring to the significance of the plant impact. The results of the ten soils samples collected in 2012 in the area show values in the same order of magnitude of the results obtained in the others characterized areas of the region and with the samples collected by the Provincial Agency of Environmental Protection in the plant valley and in the Trento surroundings during winter 2009 [19]. The limit for agricultural/residential uses has been complied with always, demonstrating that the incidence of the plant is not critical.

TABLE 2

PCDD/F soil concentrations ($ng_{I\text{-}TEQ}$/kg d.w.)	0.6 – 2.1
Values overcoming limit (agriculture/residential use)	None
PCDD/F / TEQ range	11.3 – 68.8
Distance correlation from the plant	None
Anomalies from regional data comparison	None

In order to have an overview of the role of the plant in terms of PCDD/F impact, the conveyed emissions must be taken into account too. To this concern the performed modeling demonstrated that the local incidence can be acceptable considering the characteristics of the area, but the authorized PCDD/F stream cannot be considered negligible. Thus particular attention must be put towards the stack characterizations. Coherent to this vision, the company of the plant decided to buy a continuous sampling device for average PCDD/Fs characterizations. An aspect to be optimized is the balance costbenefits of the analytical costs (analyses could be performed monthly, weekly, etc.). However, the presence of diffused emissions makes it necessary to integrate this approach with a strategy of monitoring based on deposimeters and soil sampling. Some aspects must be taken into account. When a deposimeter is placed on the ground:

- Every day weather conditions must be checked in order to calculate the cumulated rain (or snow in winter);
- In winter the air temperature must be checked daily in order to verify the risk of ice in the deposimeter;
- Every day the operating hours of the plant must be recorded as well as data on the kind of material fed;
- Once a week a visit should be scheduled to check the integrity of the device.

For the case study, the closest site (light circle, CRZ in Figure 2) can be confirmed for future characterizations, as sensible to the diffused emissions. Summer and winter depositions with the plant operating could be the base of the characterization.

A second site should be the one located at the primary school of the closest village (light circle, Borgo in Figure 2), because of the specificity of the site (presence of children; residential area).

A verification of a third site (black circle, Roncegno in Figure 2) is compulsory because of some significant anomalies in the reported data. This verification is not dependent on the steel making plant and the additional characterizations could be stopped when the problem will be solved.

Concerning soil, the garden of the primary school in Borgo could be chosen as reference site, with one characterization per year. The presented approach can be easily adapted also in other case studies where the impact of a steel making plant must be checked.

6.4 CONCLUSIONS

- The adoption of deposimeters to prevent anomalous PCDD/F impacts of a steel making plant is a viable option. However it is important to organize a continuous verification of the status of this kind of tools, in order to prevent flooding, icing, etc.

- The present case study demonstrated that crossing modelling and preliminary deposimeters characterizations offers a powerful tool to optimize this
  kind of unconventional monitoring. From another point of view, soil characterization can support this monitoring generating data useful but with the
  limit to average the effects of the impact along the years.
- Thus soil analyses cannot be the only strategy to be performed.

REFERENCES

1. Choi, S.D., Baek, S.Y., Chang, Y.S. Atmospheric levels and distribution of dioxin-
   like polychlorinated biphenyls (PCBs) and polybrominated diphenylethers (PBDEs)
   in the vicinity of an iron and steel making plant, Atmospheric Environment, 42(10),
   pp. 2479–2488, 2008.
2. Ragazzi, M., Rada, E.C., Girelli, E., Tubino, M., Tirler, W., Dioxin deposition in the
   surroundings of a sintering plant, Organohalogen Compounds, 73, pp. 1920-1923,
   2011.
3. Ames, M, Zemba, S., Green, L., Botelho, M.J., Gossman, D., Linkov, I., Palma-
   Oliveira, J., Polychlorinated dibenzo(p)dioxin and furan (PCDD/F) congener profiles in cement kiln emissions and impacts, Science of the Total Environment, 419,
   pp. 37–43, 2012.
4. Ionescu, G., Zardi, D., Tirler, W., Rada, E.C., Ragazzi, M., A critical analysis of
   emissions and atmospheric dispersion of pollutants from plants for the treatment of
   residual municipal solid waste, U.P.B. Scientific Bulletin, series C, 74(4), pp. 227-
   240, 2012.
5. Ragazzi, M., Rada, E.C. Multi-step approach for comparing the local air pollution
   contributions of conventional and innovative MSW thermo-chemical treatments,
   Chemosphere, 89(6), pp. 694-701, 2012
6. Rada, E.C., Ragazzi, M., Zardi, D., Laiti, L., Ferrari, A, PCDD/F enviromental impact from municipal solid waste bio-drying plant, Chemosphere, 84(3), pp. 289-295,
   2011.
7. Schuhmacher, M., Agramunt, M.C., Bocio, A., Domingo, J.L., de Kok, H.A.M., Annual variation in the levels of metals and PCDD/PCDFs in soil and herbage samples
   collected near a cement plant, Environment International 29, pp. 415–21, 2003.
8. Aristizábal, B., González, C., Morales, L., Abalos, M., Abad, E., Polychlorinated
   dibenzo-p-dioxin and dibenzofuran in urban air of an Andean city, Chemosphere
   85(2), pp. 170–178, 2011.
9. Weber, R., Watson, A., Assessment of the PCDD/F fingerprint of the dioxin food
   scandal from bio-diesel in Germany and possible PCDD/F sources, Organohalogen
   Compounds, 73, pp. 400-403, 2011.
10. Ryan, J.J., Rawn, D.F.K., Polychlorinated dioxins, furans (PCDD/Fs), and polychlo-
    rinated biphenyls (PCBs) and their trends in Canadian human milk from 1992 to
    2005, Chemosphere, 102, pp. 76-86, 2014.

11. Wittsiepe, J., Fürst, P., Schrey, P., Lemm, F., Kraft, M., Eberwein, G., Winneke, G., Wilhelm, M., PCDD/F and dioxin-like PCB in human blood and milk from German mothers, Chemosphere, 67(9), pp. S286-S294,2007.

12. Chang, Y.M., Fan, W.P., Dai, W.C., His, H.C., Wu, C.H., Che, C.H., Characteristics of PCDD/F content in fly ash discharged from municipal solid waste incinerators, Journal of Hazardous Materials, 192(2), pp. 521-529, 2011.

13. Rada, E.C., Schiavon, M., Ragazzi, M., Seeking potentially anomalous human exposures to PCDD/Fs PCBs through sewage sludge characterization, Bioremediation and Biodegradation Journal, 5(8), pp.1-10, 2013.

14. Schiavon, M., Ragazzi, M., Rada, E.C, A proposal for a diet-based local PCDD/F deposition limit, Chemosphere, 93(8), pp. 1639-1645, 2013.

15. Schecter, A.J., Colacino, J.A., Birnbaum, L.S., Dioxins: Health Effects, Environmental Health Perspectives, pp. 93–101, 2011.

16. Ragazzi, M., Rada, E.C., Tubino, M., Marconi, M., Chistè, A., Girelli, E., Deposition near a sintering plant: preliminary comparison between two methods of measurements by deposimeters, U.P.B. Scientific Bulletin, series D, 74(1), pp. 205-210, 2012.

17. Ragazzi, M., Rada, E.C., Marconi, M., Chiste, A., Fedrizzi, S., Segatta, G., Schiavon, M., Ionescu, G., Characterization of the PCDD/F in the Province of Trento, Proceedings of TMREES14 - International Conference on Technologies and Materials for Renewable Energy, Environment and Sustainability, 10-13 April 2014, Beirut, Libano.

18. Rada, E.C., Ragazzi, M., Chistè, A., Schiavon, M., Tirler, W., Tubino, M., Antonacci, G., Todeschini, I., Toffolon, M., A contribution to the evolution of the BAT concept in the steelmaking sector, Proceedings of the Sustainable Technology for Environmental Protection, pp. 1-8, 26-29 june 2012 Milano, Italy.

19. DICAM, Multidisciplinary study on the environmental impact of the steel making plant of Borgo Valsugana, 2013 (http://www.appa.provincia.tn.it/news/pagina199.html)

20. Rada, E.C., Ragazzi, M., Malloci, E., Role of levoglucosan as a tracer of wood combustion in an alpine region, Environmental Technology, 33(9), pp. 989–994, 2012.

CHAPTER 7

Deposition Near a Sintering Plant: Preliminary Comparison Between Two Methods of Measurements by Deposimeters

MARCO RAGAZZI, ELENA CRISTINA RADA, MARCO TUBINO, MARIANNA MARCONI, ALESSANDRO CHISTÈ, AND ELEONORA GIRELLI

7.1 INTRODUCTION

The sector of sintering plants, in the European Union, is considered today one of the major potential contributors to the atmospheric release of PCDD/F and consequent depositions. Anyway the emissions of a plant depend on many factors: characteristics of the input, sintering process, prevention and removal of PCDD/F from the gaseous stream. The local impact of a plant depends also on the variability of release through dif-

Deposition Near a Sintering Plant: Preliminary Comparison Between Two Methods of Measurements by Deposimeters. © *Ragazzi M, Rada EC, Tubino M, Marconi M, Chistè A, and Girelli E.* UPB Scientific Bulletin, Series D: Mechanical Engineering *74,1 (2012), http://www.scientificbulletin.upb.ro/rev_docs_arhiva/rezfa7_516263.pdf. Reprinted with permission from the authors.*

fused emissions, secondary emissions and conveyed gases, as each stream can be related to a different way of dilution into the atmosphere. The high variability of parameters that can affect the local impact from a sintering plant makes it interesting the adoption of measuring instruments, such as deposimeters, in order to contribute to the understanding of the human exposure in the surrounding area.

The present paper refers to an Italian case-study whose PCDD/F deposition measurements are in progress using two types of deposimeters placed in a selected site. Preliminary and expected values are discussed taking into account seasonality and operativity of the plant [1].

7.2 MATERIALS AND METHODS

The choice of the site for deposimeter positioning resulted from the adoption of an atmospheric diffusion model [2] specifically implemented and the analysis of the territory in order to select an area potentially exposed to the plant emissions.

A vetropyrex Depobulk® deposimeter [3] with its body in glass (Fig. 1) was selected, able to capture the overall deposition of PCDD/F generally in a period of one month around. A special ring avoids interferences from bird excreta.

In the same position, called "far", the second kind of deposimeter was installed: two vessels for collecting dry and wet depositions (Fig. 2) [4] characterize this particular technology. The vessels work alternatively "in absence" (dry) or "in presence" (wet) of atmospheric rainfall with the aid of an atmospheric rainfall sensor located on a mobile cover.

An additional Depobulk® device was positioned in a site selected thanks to the diffusion model results in order to check the incidence of the sintering plant form diffuse emissions. This site has been called "close".

The first period of parallel sampling was organized from March 2011 to June 2011, synchronizing the deposition characterization with the activity of the plant (that stopped 5 weeks for the summer closure in August-September). An accurate cleaning was planned each time when the deposimeters had to operate. Rainy periods were observed in details in order to avoid an overflow of the collected water for the conventional deposimeter (data were taken from a local meteorological station). Each amount of PCDD/F measured in the collected mass was divided by the surface of the deposimeter and the period of characterisation.

The "far" site chosen for the PCDD/F deposition characterization is inside the area of a primary school coupled with a kindergarten. The reason of the choice was related to the location of the school, generally downwind the sintering plant when the plant is operating, to the proximity to the emission source, to the sensibility of the receptor (children) and to the demographic density of the area.

The "close" site was chosen at the limit of the private area of the sintering plant. For reasons related to the cost of electricity, the plant operates from 8 p.m. to 8 a.m. during weekly days and 24 h d^{-1} during the weekend [1].

7.3 PRELIMINARY RESULTS

The first step of the activities was the verification, by diffusion modeling, of the area of interest for the PCDD/F depositions. The "far" site shown in Fig. 3 was selected as representative both of the depositions from the stack and of the diffused emissions (not conveyed). The "close" site of Fig. 3 was selected as representative mainly of the diffused emissions incidence. The first is at a distance of around 1400 m (from the stack) and the second at a distance of around 700 m (from the stack; the plant buildings develop from the stack towards the village).

Generally the influence of a plant on the PCDD/F concentration in the surrounding area could be detected by measuring higher values of deposition close to the installation. This is what was detected in the case-study, as shown in Fig 4, for three different periods of the year in spring and early summer:

- Period 1: 36 days, month : March - April 2011;
- Period 2: 40 days, month : April - May 2011;
- Period 3: 41 days, month : May - June 2011.

FIGURE 1: Conventional deposimeter: before installation (left) and on the roof (right).

FIGURE 2: Wet&dry deposimeter on the roof of the selected building

FIGURE 3: Selected positions for the deposimeters (close and far from the plant).

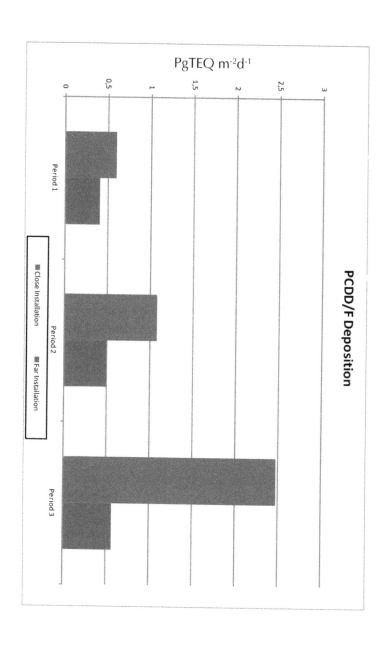

FIGURE 4: Values of deposition close and far from the plant.

These values remained below some limits available in the literature: in the Flanders a range of 3.4–14 pg_{TEQ} m^{-2} d^{-1} is proposed [5] while in Germany a value of 15 pg_{TEQ} m^{-2} d^{-1} is taken into account [6]. The values found are similar to of lower than the ones measured in rural areas: in Denmark a range of 2.2–3.3 pg_{TEQ} m^{-2} d^{-1} was found in rural forest sites [7].

From the graph in Fig. 4 it is clear that the plant may have some influence on air pollution of the surrounding area but a deeper analysis must be made to verify its role. To this concern, a further comparison of analyses made in two distinct periods (period of operation and period of inactivity of the plant for holiday reasons) will help to understand its role.

These analyses were planned but results are not yet available. Also the preliminary results from wet & dry deposimeters are not yet available but some considerations can be made concerning the expected results. Indeed these low values of deposition can create some problems in the analyses.

The main problem can be identified in the detection limit of the instruments used to perform the analyses. Dividing the depositions on two devices (vessels) it could happen that the concentration values for a part of the dioxins and a part of the furans are each below the detection limit with an occurrence higher than the one resulting from conventional deposimeters. To this concern it is interesting to carry out a preliminary check to understand which apportionment between wet & dry deposition can cause a drop in concentrations below the detection limit. With reference to the detection limit of the analytical method of EPA 1613 B 1994 PCDD/F of 10 fg sample^{-1}, it is interesting to see when the division between wet & dry deposition lowers the value of concentration below this limit.

With reference to the area described above, analyses of three different depositions of about 40 days in a PCDD/F deposimeter Depobulk® were used as starting point [8].

For each analysis an hypothetical distribution was calculated between wet & dry deposition, with the aim to find the percentage for which there is a concentration value below the limit of detection LOD (10 fg sample^{-1}) called "percentage limit obtained" (PLO) (Tab. 1).

The distributions are assumed from 100% dry to 5% dry (95% wet), calculated in steps of 5%.

TABLE 1: Percentage Limit Obtained values—PLO—for three analyses

	Period 1		Period 2		Period 3	
	Conc. (pg sample^{-1})	PLO	Conc. (pg sample^{-1})	PLO	Conc. (pg sample^{-1})	PLO
2,3,7,8 TCDD	<l.o.d.	100	0.010	95	<l.o.d.	100
1,2,3,7,8 PeCDD	0.012	80	0.443	NR	0.021	45
1,2,3,4,7,8 HxCDD	0.047	20	0.016	60	0.032	30
1,2,3,6,7,8 HxCDD	0.580	NR	0.099	10	0.091	10
1,2,3,7,8,9 HxCDD	0.096	10	0.031	30	0.047	20
1,2,3,4,6,7,8 HpCDD	1.130	NR	4.733	NR	2.520	NR
OCDD	13.500	NR	21.260	NR	9.380	NR
2,3,7,8 TCDF	1.040	NR	1.105	NR	0.494	NR
1,2,3,7,8 PeCDF	0.357	NR	0.116	5	0.162	5
2,3,4,7,8 PeCDF	0.015	65	0.139	5	0.443	NR
1,2,3,4,7,8 HxCDF	0.224	NR	0.138	5	1.010	NR
1,2,3,6,7,8 HxCDF	0.238	NR	0.279	NR	0.950	NR
2,3,4,6,7,8 HxCDF	1.090	NR	<l.o.d	100	1.190	NR
1,2,3,7,8,9 HxCDF	0.017	55	0.016	60	0.016	60
1,2,3,4,6,7,8 HpCDF	3.820	NR	2.683	NR	4.560	NR
1,2,3,4,7,8,9 HpCDF	0.395	NR	0.044	20	0.228	NR
OCDF	0.880	NR	0.620	NR	4.240	NR

In the favorable event that the limit of detection is not reached with any distribution percentage, NR (Not Reached) is shown in the tables. The combination of these values can cause some problems in obtaining a complete fingerprint of the depositions if the distribution wet & dry is not favorable.

7.6 CONCLUSIONS

Conventional deposimeters can generate important information in order to study the incidence of a plant. Additional information can come from wet & dry devices, to be used for a deeper dispersion modeling analysis.

Preliminary analyses of expected data from wet & dry deposimeters can halp in choosing the lasting of the sampling.

REFERENCES

1. M. Ragazzi, E.C. Rada, E. Girelli, M. ,Tubino, W. Tirler, Dioxin deposition in the surroundings of a sintering plant, Proceedings of 31st International Symposium on Halogenated Persistent Organic Pollutants POPs' Science in the Hearth of Europe, Brussels Belgium, Dioxin 2011, Paper 4121.
2. AERMOD model, version 09292 with pre-processor AERMET and AERMAP, U.S. EPA distribution, 2009.
3. http://www.labservice.it/index.php?keyword=depo&categoria=Ambientale&sottoc ategoria=&bSearch=Cerca&x=strumento&id_strumento=LSDEPOLABGL
4. Wet & Dry deposimetr, FAS005AB mode, 2011. http://www.mtx.it/ita/strumenti. html.
5. L. Van Lieshout, M. Desmedt, E. Roekens, R. De Frè, R. Van Cleuvenbergen, M. Wevers, Deposition of dioxins in Flanders (Belgium) and a proposition for guide values, in Atmospheric Environment, vol. 35/1, 2001, pp. 383-390.
6. LAI, Laenderausschuss fuer Immissiosschutz (State Committee for Pollution Control, 2000;
7. M.F. Hofmand, J. Vikelsoe, H.V. Andersen, Atmospheric bulk deposition of dioxin and furans to Danish background areas, in Atmospheric Environment, vol. 41/11, 2007, pp. 2400-2411.
8. Eco-research, Analytical data, 2011.

CHAPTER 8

Estimation of Regional Air-Quality Damages from Marcellus Shale Natural Gas Extraction in Pennsylvania

AVIVA LITOVITZ, AIMEE CURTRIGHT, SHMUEL ABRAMZON, NICHOLAS BURGER, AND CONSTANTINE SAMARAS

8.1 INTRODUCTION

Recent technological innovations in natural gas extraction—namely the combined use of horizontal drilling and hydraulic fracturing—are enabling access to vast new natural gas resources contained in shale deposits across the United States (Kargbo et al 2010, Mooney 2011). The Marcellus Shale formation is the largest US shale gas deposit and has contributed significantly in recent years to increased US natural gas production (US DOE EIA 2012a, 2012b). The rapid development of this resource has been touted as both an economic boon (Considine et al 2011, Marcellus Shale Coalition 2012) and a potential environmental mistake for the region (PennEnvironment Research and Policy Center 2012). Environmental concerns often relate to risks to water resources (Ground Water Protection Council and ALL Consulting 2009, Mooney 2011). However, utilizing natural gas from shale deposits also produces air emissions of various types during extraction, transportation, and end use.

Increases in conventional air pollution may pose a threat to air-quality in shale gas extraction regions (Shogren 2011, Alvarez and Paranhos 2012, McKenzie et al 2012, Steinzor et al 2012). Such emissions can have direct physical impacts on health, infrastructure, agriculture and ecosystems. For example, short-term exposure to criteria pollutants such as sulfur dioxide (SO_2) and nitrogen oxides (NOx) has been linked to adverse respiratory effects. Exposure to fine particulate matter (PM) and ozone (O_3) may increase respiratory-related hospital admissions, emergency room visits, and premature death. The expanded use of natural gas could arguably reduce net emissions from the electricity sector if used in lieu of coal (US EPA 1999, NRC 2010)[1]. However, shale gas extraction activities such as diesel truck transport and natural gas processing at compressor stations could lead to increases in air pollution in regions where extraction occurs.

Life cycle greenhouse gas (GHG) emissions from shale gas are often assessed to be greater than conventional natural gas. However, most studies also indicate that expanded use of shale gas could lower net GHG emissions relative to coal-based electricity (Burnham et al 2011, Fulton et al 2011, Hultman et al 2011, Jiang et al 2011, Venkatesh et al 2011, Lu et al 2012, Skone et al 2012, Weber and Clavin 2012). Additionally, any GHG benefits from shale gas use are not localized to the region where extraction occurs. While GHGs are an important consideration, this letter focuses on conventional, non-GHG air pollution.

A recent GAO literature survey found evidence that extraction activities pose risks to air quality. While some studies indicated degraded air quality at specific shale gas extraction sites, the data necessary to quantify aggregate impacts were not available (US Government Accountability Office 2012). Pennsylvania recently mandated reporting on some emissions to the Pennsylvania Department of Environmental Protection (PA DEP), but this data collection has just begun (Pennsylvania Department of Environmental Protection 2011). This analysis provides initial, first-order estimates of regional air emissions generated by Pennsylvania-based extraction activities[2] and associated ranges of potential regional monetized damages. These estimates must be considered in the context of other external costs and benefits of shale gas extraction and use, and should be refined as new data becomes available.

8.2 ESTIMATING LOCAL EMISSIONS AND REGIONAL DAMAGE FROM SHALE GAS EXTRACTION ACTIVITIES

The major stages of shale gas extraction considered here are depicted in figure 1, and emissions occur across many of them (NYS DEC 2011). This analysis includes emissions associated with four shale gas-related activities:

- Diesel and road dust emissions from trucks transporting water and equipment to the site, and wastewater away (stages 2 and 8 in figure 1);
- Emissions from well drilling and hydraulic fracturing, including diesel combustion (stage 4);
- Emissions from the production of natural gas, including on-site diesel combustion and fugitive emissions (stage 5);
- Combustion emissions from natural gas powered compressor stations (stage 7).

We omit emissions from venting or flaring at well-sites (stages 4 and 5). The US EPA will prohibit this by 2015, requiring so-called "green completions" which capture completions emissions rather than venting or flaring them (United States Environmental Protection Agency 2012), and many natural gas producers have already begun following this practice. Industry-reported emissions for venting are small relative to other sources; however flaring-emission estimates may have a more substantial impact[3].

Pollutants assessed were: volatile organic compounds (VOCs)[4]; NOx; PM_{10} (<10 μm); $PM_{2.5}$ (<2.5 μm);[5] and SO_2.[6] We focus on these due to their adverse impacts and regulatory status; accordingly, they often appear in facility permitting and emissions reporting, and all are included in the model used here to monetize damages. Table 1 summarizes air pollutants and extraction activities included in this analysis.

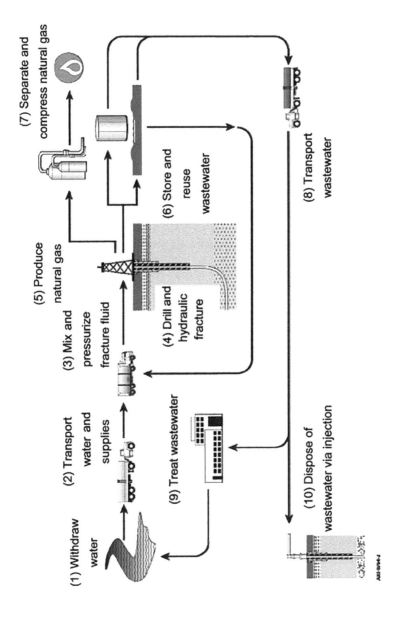

FIGURE 1: Major activities of shale gas extraction using horizontal drilling and hydraulic fracturing.

TABLE 1: Air emissions damages, localization of effects, and relevant pollutants of concern.

Damage category	Damage location	Relevant emissions	Relevant stages	Inclusion in this analysis
Climate change	Local, regional, and global	GHGs: CO_2, CH_4, N_2O, O_3	• Stages 2, 8: transport	No GHGs included in this study
			• Stages 3, 4, 5: site activities	
			• Stage 7: processing	
Air quality	Local and regional	VOCs, NOx, PM, SO_2, O_3, CO	• Stages 2, 8: transport	Development activities: (1) transport; (2) well drilling, hydraulic fracturing
			• Stages 3, 4, 5: site activities	
			• Stage 6: wastewater storage and reuse	Ongoing activities: (3) production; (4) compressor stations
			• Stage 7: processing	Pollutants: direct: VOCs, NOx, PM, SO_2; indirect: O_3 via VOCs and NOx

8.3 METHODS USED TO CALCULATE AIR POLLUTION DAMAGES

There is considerable uncertainty in emissions associated with shale gas development. This is due to a scarcity of emissions data and to actual differences in emissions caused by regional and site-specific variations in technology and processes[7]. The several estimation methods and data sources we use result in a wide range of estimates. For industry data used here, estimation methods are likely to have been used (e.g., an emissions factor approach) rather than empirical determinations. Such estimations often differ widely from empirical findings, especially for fugitive emissions (Chambers et al 2008, Pétron et al 2012), which are also subject to uncertainty (Levi 2012).

Our approach to estimating regional air pollution damages is modeled after another study of the external costs of energy production (NRC 2010). For each activity we have estimated emissions on a per well or per-unit-of-natural-gas-produced basis. Compressor station emissions are estimated per station. These emissions estimates allow us to obtain total statewide emissions, with resolution at the county-level, that we convert to statewide damages using the Air Pollution Emission Experiments and Policy (AP-EEP) model (Muller and Mendelsohn 2007, 2012). We first describe our approach for estimating emissions (sections 3.1–3.5) and then describe how these emissions were converted into monetary damages (section 3.6).

TABLE 2: Range of assumed well-site development emissions in this analysis.

Emissions activity	VOC	NOx	PM2.5	PM10	SOx
Total diesel and road dust development emissions (kg/well)	18–31	320–580	9.4–32[a]	9.8–32[a]	0.47–0.79
Total well-site development emissions (kg/well)	150–170	3800–4600	87–130	87–130[b]	3.8–110

[a]PM_{10} emissions were unavailable for heavy-duty trucks; in this case, it was assumed all diesel-related PM emissions were less than 2.5 μm. All road dust was also assumed less than 2.5 μm. Therefore aggregate PM_{10} counts differ from $PM_{2.5}$ only in light-duty vehicle emissions; at the high end of our range, this difference is not significant. [b]Industry reporting often assumes all PM emissions are less than 2.5 μm and so PM_{10} counts are almost the same as $PM_{2.5}$.

10.3.1 ESTIMATES OF AIR POLLUTANT EMISSIONS FROM TRANSPORT TRUCKS

Diesel trucks used to transport water and supplies to and from the well-site emit air pollutants. Our assumption of the total number of per well truck trips is based on the New York State Department of Environmental Conservation's (NYS DEC) 2011 Environmental Impact Statement (EIS) (NYS DEC 2011). The corresponding implied diesel emissions were estimated with emissions factors in the Greenhouse gases, Regulated Emissions, and Energy use in Transportation (GREET) model (US DOE Argonne National Labs (ANL) 2012) and in a recent National Research Council study (NRC 2010) for light-duty and heavy-duty vehicles, respectively. Truck traffic can also result in considerable road dust, which we include based on estimates in the NYS EIS. Additional details are provided in section S.1 (available at stacks.iop.org/ERL/8/014017/mmedia). Table 2 provides the total per well transport emissions assumed.

8.3.2 ESTIMATES OF ON-SITE AIR POLLUTANT EMISSIONS FROM WELL CONSTRUCTION

Well development generates emissions at the extraction site during well pad construction, drilling, and hydraulic fracturing. The range of well-site construction emissions used in this analysis were estimated using data reported by three major regional shale gas producers, including one set of emissions reported directly to us and two sets obtained through PA DEP as part of its Air Emissions Inventory for the Natural Gas Industry (PA DEP 2011, Pennsylvania Department of Environmental Protection 2011, Ramamurthy 2012). Details on these data sets and how they were used are provided in section S.2 (available at stacks.iop.org/ERL/8/014017/mmedia); final values used in this analysis are provided in table 2.

TABLE 3: Range of assumed well-site production emissions used in this analysis.

Emissions activity	VOC	NOx	$PM_{2.5}$	PM_{10}	SOx
Total annual well-site production emissions per well (kg/well)	46–1200	520–660	9.9–50	9.9–50[a]	3.1–4.0

[a]*Industry reporting often assumes all PM emissions are less than 2.5 μm and so PM_{10} counts are here the same as $PM_{2.5}$.*

8.3.3 ESTIMATES OF AIR POLLUTANT EMISSIONS FROM SHALE GAS PRODUCTION

The ongoing production of shale gas also generates emissions. Data were obtained from two major regional operators and were used to establish low and high values of production emissions estimates, shown in table 3. Production emissions obtained for this analysis were less consistent between sources than construction emissions, although values are typically within an order of magnitude. In addition to differences between producers, this range may also reflect differences in the operators' reporting assumptions (see section S.3 available at stacks.iop.org/ERL/8/014017/mmedia).

8.3.4 ESTIMATES OF AIR POLLUTANT EMISSIONS FROM COMPRESSOR STATIONS

Emissions from compressor stations continue over the long term as natural gas is produced over the life of many wells. To estimate ranges of potential emissions from compressor stations, we reviewed permit applications for more than a dozen new facilities permitted in Pennsylvania in 2010 and 2011, as described in section S.4 (available at stacks.iop.org/ERL/8/014017/mmedia). We make use of the facility-wide potential-to-emit (PTE) emissions values, with ranges reflecting the lows and highs observed in our review. If most facilities are operating below capacity, they may fall at the lower end of the estimate; on the other hand, if they are not running optimally (e.g., frequent shut-downs and start-ups), the emissions could be even higher than indicated by PTE. Values in table 4 therefore represent a range of operating situations.

TABLE 4: Range of compressor station emissions estimates used in this analysis.

Emissions activity	VOC	NOx	$PM_{2.5}$	PM_{10}	SOx
Total annual compressor station emissions (metric tons/facility)	11–45	46–90	1.4–5.5	1.4–5.5[a]	0–1.7

[a]*Industry reporting often assumes all PM emissions are less than 2.5 µm and so PM10 counts are here the same as PM2.5.*

8.3.5 AGGREGATED AIR POLLUTANT EMISSIONS ESTIMATES

We used per-facility emissions to estimate county-level and statewide emissions. We present total statewide aggregated emissions in table 5. These values represent the ranges of emissions in tables 2–4 applied to the following extraction activity assumptions for 2011: construction of 1741 wells; statewide shale gas production of nearly 1.1 trillion cubic feet; and operation of 200 recently developed compressor stations. County-level assumptions and values can be found in section S.5 (available at stacks.iop. org/ERL/8/014017/mmedia).

TABLE 5: Statewide emissions estimates for shale gas development and production in 2011.

Activities	Statewide annual emissions (metric tons per year)				
	VOC	NOx	$PM_{2.5}$	PM_{10}	SOx
(1) Transport	31–54	550–1000	16–30	17–30	0.82–1.4
(2) Well drilling and hydraulic fracturing	260–290	6600–8100	150–220	150–220	6.6–190
(3) Production	71–1800	810–1000	15–78	15–78	4.8–6.2
(4) Compressor stations	2200–8900	9300–18 000	280–1100	280–1100	0–340
Total[a]	2500–11 000	17 000–28 000	460–1400	460–1400	12–540

[a]*These totals are reported to two significant figures, as are all intermediate emissions values in this document. The activity emissions may not exactly sum to the totals.*

8.3.6 ESTIMATING DAMAGES FROM AIR POLLUTANT EMISSIONS

For each of the four activities included in this analysis, emissions per well or per million cubic feet were used to estimate county-level emissions because damage per unit of pollution varies greatly with location. These county-level emissions were then converted into county-level annual damages using the APEEP model (Muller and Mendelsohn 2007, 2012). APEEP is an integrated assessment model that uses information derived from the air quality and epidemiological literature[8]. APEEP converts tons of pollutant emitted into physical health and environmental damages, including mortality, morbidity, crop and timber loss, visibility, and effects on anthropogenic structures and natural ecosystems. The base APEEP model calculates age-specific health damages, recognizing that mortality risk and lost years of life will vary with age. Section S.6 (available at stacks.iop.

org/ERL/8/014017/mmedia) provides additional details and damages for each county. The damage ranges given for each county are a result of the ranges in emissions estimates above; in addition, because of uncertainty in the size of PM, for activities 2–4 the low damage estimates assume none of the PM is $PM_{2.5}$ and the high damage estimates assume that all PM is $PM_{2.5}$. Complete damages by county and pollutant are found in tables S.11 and S.12 (available at stacks.iop.org/ERL/8/014017/mmedia).

8.4 RESULTS

8.4.1 REGIONAL SHALE EXTRACTION AIR POLLUTANT DAMAGE ESTIMATES

The aggregated estimated regional damages associated with Pennsylvania shale gas extraction activities are shown in table 6. The total regional air-quality-related damages, at the level of development and production in Pennsylvania in 2011, ranged between $7.2 million and $32 million. These represent the sum of damages in all Pennsylvania counties. While per unit damages will vary greatly with location of the emissions, we also calculated the average per well or per MMCF damages. Some extraction activities occur in regions of Pennsylvania that influence the air quality of populated areas of other states; so while our estimates of emissions were confined to extraction activities in the state of Pennsylvania, these damages should be considered a regional impact, given that pollutants may cross the state border.

Development activities represent about a third or less of total extraction-related emissions (35–17% across the estimated range), whereas ongoing activities represent the majority of emissions (65–83% across the range). Compressor station activities alone represent 60–75% of all extraction-associated damages. Considering the relative importance of different pollutants, VOCs, NOx, and $PM_{2.5}$ combined across all activities were responsible for 94% of total damages; across the range of estimates they contributed 34–33%, 59–20%, and 2–41%, respectively (shown by activity in table S.11 at stacks.iop.org/ERL/8/014017/mmedia).

TABLE 6: Estimates of regional air pollution damages from Pennsylvania extraction activities in 2011.

Activities	Timeframe	Total regional damage for 2011 ($2011)	Average per well or per MMCF damage ($2011)
(1) Transport	Development	$320 000–$810 000	$180–$460 per well
(2) Well drilling, fracturing	Development	$2 200 000–$4 700 000	$1 200–$2 700 per well
(3) Production	Ongoing	$290 000–$2 700 000	$0.27–$2.60 per MMCF
(4) Compressor stations	Ongoing	$4 400 000–$24 000 000	$4.20–$23.00 per MMCF
(1)–(4) Aggregated	Both	$7 200 000–$32 000 000	NA

8.4.2 COMPARISON OF AIR POLLUTANT EMISSIONS AND DAMAGES TO OTHER INDUSTRIAL SECTORS IN PENNSYLVANIA

To assess the relative impact the shale gas industry might have on regional air quality, we compare the total emissions estimated for extraction activities in 2011 with net emissions from other major sectors of the Pennsylvania economy. We obtained data from the US EPA's 2008 National Emissions Inventory (NEI) (US EPA 2008) and calculated statewide emissions (see section S.7 available at stacks.iop.org/ERL/8/014017/mmedia). These statewide totals are presented in table 7, along with the percentage of these total emissions that shale gas extraction activities in 2011 represent. Compared to total emissions from all industries reporting, the shale extraction industry in 2011 was producing relatively little conventional air pollution. Only NOx emissions are equivalent to more than 1% of statewide emissions across the entire estimated range.

Extraction activities, however, are not evenly distributed throughout the state, so it is instructive to look at the magnitude of emissions in the few counties where activities were concentrated in 2011. More than 20% of wells were found in one county and nearly 50% were in the top 3 counties; the 10 counties with the most development constituted nearly 90% of wells in the state (see table S.8 available at stacks.iop.org/ERL/8/014017/

mmedia). The statewide extraction industry also produced VOC[9] and NOx[10] emissions equivalent to or larger than some of the largest single emitters in the state—GW-scale coal-based electric power plants. In the counties with the most activity, even the low-end of the NOx emissions estimate ranges were 20–40 times higher than the level that would constitute a "major" emissions source, although individually the new shale-related facilities are generally not subject to major source permit requirements. On the other hand, the magnitude of PM and SO_2 emissions are much less significant relative to existing major sources, as the statewide totals imply[11].

TABLE 7: Magnitude of shale gas extraction industry relative to air pollutant emissions from other industrial sectors in Pennsylvania.

Total sector or comparison	VOCs	NOx	$PM_{2.5}$	PM_{10}	SOx
Shale gas extraction industry in 2011, from table 5 (metric tons)	2500–11 000	17 000–28 000	460–1400	460–1400	12–540
Total from EPA/ NEI, all sectors reporting (metric tons)[a]	720 000	579 000	134 000	322 000	898 000
Shale extraction relative to total (%)	0.35–1.5	2.9–4.8	0.34–1.0	0.14–0.43	0.0013–0.060

[a]*Combustion-based electric utilities and highway and off-highway vehicles generally constitute a large percentage of statewide emissions in EPA's 2008 NEI. For example, combustion-based electricity production, highway vehicles, and off-highway vehicles sectors statewide represent: 80% of NOx (460 000 of 580 000 metric tons); 47% of $PM_{2.5}$ (63 000 of 130 000 metric tons); and 87% of SO_2 (780 000 of 900 000 metric tons). Combined, they are less significant for VOCs and PM_{10} (26% and 22% of statewide respectively).*

Although the correlation with emissions is not direct, the total regional damages from the shale gas extraction industry are also expected to be small relative to statewide air pollution emissions damages[12]. For comparison, we estimate that the largest coal-fired power plant in Pennsylvania—while not the state's most polluting facility—alone produced about

$75 million in damages in 2008. The four largest facilities—which included the top two SO_2 emitters in the state—produced nearly $1.5 billion in damages in 2008. For the shale gas extraction industry, monetary damages were driven by significant levels of VOCs, NOx, and $PM_{2.5}$, and the whole industry constituted less than 2%, 5%, and 1% for each of the pollutants, respectively, of total emissions in the state in 2008 from all industries reporting.

Because the relative damages will tend to be larger in the counties where shale gas extraction activities are concentrated, where population is relatively high, and where air quality is already a concern, it is also important to consider the county-level damage. For example, Washington County had the fifth largest number of wells (156) in 2011 but resulted in the highest damages, estimated at $1.2–8.3 million. Damage in this county represented about 20% of statewide damages from the extraction industry[13]. And while not typical of 2011 development, this example illustrates the potential impact of extraction when located in relatively populated areas[14].

8.5 DISCUSSION

We estimate that total regional air-quality-related damages, at the level of development and production in Pennsylvania in 2011, ranged between $7.2 million and $32 million (table 6). However, extraction industry damages will not be constant over time or evenly distributed in space, and there are important policy implications of when and where emissions damages occur. Development emissions damages range from about $2.5 to $5.5 million, but the majority of annual attributable emissions will continue for the life of the well and associated compressor facilities. This is true despite the relatively high level of development activity in 2011 and the relatively low number of actively producing shale gas wells, compared to what is expected in coming years. At the low end of our estimates, 66% of total damages in 2011 were attributable to long-term activities; at the high end, more than 80% of damages occur in the years after the well is developed. Nor are most emissions associated with well-site activities. More than half of emissions damages from this industry come from compressor stations, which may serve dozens of individual wells, including conventional ones. Our estimates indicate that regulatory agencies and the shale gas industry, in developing regulations and best practices, should account for air emissions from ongoing, long-term activities and not just emissions associated with development, such as drilling and hydraulic fracturing, where much attention has been focused to date. Even if development slows in the Marcellus region, as it did in 2012, the long-term nature of these emission sources will mean that any new development will add to this baseline of emissions burden as more producing wells and compressor stations come online.

Additionally, most development activities do not constitute "major sources" under federal air-quality regulations. Especially for those counties that already suffer from high levels of air pollution (i.e., those in or near Clean Air Act non-attainment status), these new activities may make meeting federal air-quality standards more difficult. This issue was raised in the context of the Haynesville Shale region, where authors noted that emissions could "be sufficiently large that (they)...may affect the ozone attainment status" (Kemball-Cook et al 2010). It may be hard to limit these emissions through mechanisms such as permitting restrictions, which typically do not apply to mobile and minor stationary sources. Existing regulations may therefore not be well-suited for managing emissions from a substantial number of small-scale emitters. Proposals to aggregate industry sources should be carefully considered in terms of the appropriate unit of aggregation (e.g., by company, by geographic region) and any unintended consequences or perverse incentive they may create. One approach to reducing air emissions is to require the use of Best Available Technologies (BAT); for compressors, these include lean-burn engines, non-selective catalytic reduction, or electrification, measures often found to be cost-effective (Armendariz 2009). The various costs of meeting or exceeding BAT in Pennsylvania will likely be estimated to support updated compressor permit requirements in Pennsylvania in 2013.

It is worth stressing that a substantial portion of emissions estimated here are not specifically attributable to the "unconventional" nature of shale gas. Natural gas compressor stations are necessary to produce and distribute natural gas from any source, from conventional to biomethane. So while the emissions levels estimated are non-trivial, they may not differ substantially from any other large-scale industrial emissions that impact regional air quality; it is the scale of the resource extraction or industrial activity that is likely to matter most. Additionally, the magnitude of the potential damages must be considered in the context of other external costs associated with this industry, as well as in terms of the potential benefits of shale gas use.

While statewide emissions from the extraction industry are relatively small compared to some other major sources of air pollution in the state (e.g., SO_2 from GW-scale coal-fired power plants), these emissions sources are nevertheless a concern in regions of significant extraction activities. More detailed analyses, including regional data acquisition and consideration of site-specific variability, will be valuable in regions of intense extraction activity and for specific activities and pollutants shown in this analysis to be of most potential concern. And while significant uncertainty may exist for some potential risks of shale gas extraction, under current standard practices, shale gas extraction will be associated with non-trivial air pollution emissions.

FOOTNOTES

1. Emissions relative to renewable technologies are generally estimated to be lower than those of natural gas, so using natural gas in lieu of renewables would increase emissions.

2. This analysis does not specifically address acute damages resulting from short-term, high levels of exposure near well-sites but rather focuses on region-wide damages from a general degradation in air quality.

3. For industry inventories that report venting, these emissions are less than 0.1% of VOCs from well drilling and hydraulic fracturing, as described in section 3.2. However, another source (NYS DEC 2011) estimates that total drilling, fracturing, and production PM emissions increase by 250% with flaring; NOx and VOCs increase by 120%. Assuming these increases, and that all wells flare completions emissions and all PM from flaring is $PM_{2.5}$, additional damages are $5.7 million, or 18% of our high-bound total damage estimate.

4. The EPA defines VOCs to include organic compounds that undergo photochemical reactions in the atmosphere and does not include methane.

5. PM_{10} typically includes all particles less than 10 μm and $PM_{2.5}$ all particles less than 2.5 μm. Thus PM_{10} includes $PM_{2.5}$ in most reporting. In industry reports, there is considerable uncertainty in PM size, and it is often assumed that all PM is smaller

than 2.5 µm (i.e. $PM_{10} = PM_{2.5}$). $PM_{2.5}$ has much larger health effects than PM_{10}; this assumption therefore implies the maximal damage.

6. In some cases sulfur oxides are reported as a mixture (SOx); in our damage calculations, we treat all SOx as SO_2.

7. In addition to differences in practices and technologies, well-specific variables that may influence emissions include length of well bore, number of fracturing stages, geographic location, and characteristics of the natural gas formation (e.g. wet or dry gas). For emissions reported by industry, we have little knowledge of estimation methodology.

8. In considering the annual benefits of the Clean Air Act in 2000, APEEP gives a result of $48 billion compared to the US EPA's estimate of $71 billion. Muller and Mendelsohn argue that the US EPA work likely overstates benefits as it relies on air quality monitoring at sites that were out of attainment, sites likely to show greater changes in pollution levels than the country at large (2007). Note that in making the APEEP estimates, Muller and Mendelsohn use the US EPA's assumptions on value of a statistical life and concentration response function.

9. The top five and top twenty VOC emitters produce 252 metric tons per year and 542 tons per year, respectively, in 2008.

10. For example, the range of estimates of emissions of NOx is comparable to or larger than the emissions of the top four NOx emitters in the state. These top four facilities reported emissions of about: 23 500; 22 200; 16 200; and 15 800 metric tons per year of NOx. The facilities are 2.7, 1.7, 2.0, and 1.9 GW coal-fired power facilities, respectively.

11. For example, the top four emitters of SO_2 in the state produce from 90 000 to 170 000 metric tons each, so even the high end of the estimates of SO_2 for the extraction industry are equivalent to less than a per cent of these.

12. Calculation of the statewide damages of all major emitters involves estimating damages for each source individually, due to county-to-county variability of the damage function as well as accounting for each emissions source location and height, and is out of scope for this analysis.

13. These damages were equivalent to about 11% of the damages from the largest electricity plant.

14. In this case, Washington County is just south of Allegheny County and the city of Pittsburgh; previous development in the state occurred in more rural north and central Pennsylvania.

REFERENCES

1. Alvarez R A and Paranhos E 2012 Air pollution issues associated with natural gas and oil operations EM pp 22–5 (www.edf.org/sites/default/files/AWMA-EM-airPollutionFromOilAndGas.pdf)

2. Armendariz A 2009 Emissions from Natural Gas Production in the Barnett Shale Area and Opportunities for Cost-Effective Improvements (Austin, TX: Environ-

mental Defense Fund) (www.edf.org/sites/default/files/9235_Barnett_Shale_Report.pdf)

3. Burnham A, Han J, Clark C E, Wang M, Dunn J B and Palou-Rivera I 2011 Life-cycle greenhouse gas emissions of shale gas, natural gas, coal, and petroleum Environ. Sci. Technol. 46 619–27

4. Chambers A K, Strosher M, Wootton T, Moncrieff J and McCready P 2008 Direct measurement of fugitive emissions of hydrocarbons from a refinery J. Air Waste Manag. Assoc. 58 1047–56

5. Considine T J, Watson R and Blumsack S 2011 The Pennsylvania Marcellus Natural Gas Industry: Status, Economic Impacts and Future Potential (University Park, PA: Pennsylvania State University, College of Earth and Mineral Sciences, Department of Energy and Mineral Engineering)

6. Fulton M, Mellquist N, Kitasei S and Bluestein J 2011 Comparing Life-Cycle Greenhouse Gas Emissions from Natural Gas and Coal (Frankfurt: Deutsche Bank and Worldwatch Institute)

7. Ground Water Protection Council and ALL Consulting 2009 Modern Shale Gas Development in the United States: A Primer (Oklahoma City, OK: US Department of Energy Office of Fossil Energy and National Energy Technology Laboratory)

8. Hultman N, Rebois D, Scholten M and Ramig C 2011 The greenhouse impact of unconventional gas for electricity generation Environ. Res. Lett. 6 044008

9. Jiang M, Griffin W M, Hendrickson C, Jaramillo P, VanBriesen J and Venkatesh A 2011 Life cycle greenhouse gas emissions of Marcellus shale gas Environ. Res. Lett. 6 034014

10. Kargbo D M, Wilhelm R G and Campbell D J 2010 Natural gas plays in the Marcellus shale: challenges and potential opportunities Environ. Sci. Technol. 44 5679–84

11. Kemball-Cook S, Bar-Ilan A, Grant J, Parker L, Jung J, Santamaria W, Mathews J and Yarwood G 2010 Ozone impacts of natural gas development in the haynesville shale Environ. Sci. Technol. 44 9357–63

12. Levi M A 2012 Comment on 'Hydrocarbon emissions characterization in the Colorado Front Range: a pilot study' by Gabrielle Pétron et al J. Geophys. Res. 117 D21203

13. Lu X, Salovaara J and McElroy M B 2012 Implications of the recent reductions in natural gas prices for emissions of CO2 from the US Power Sector Environ. Sci. Technol. 46 3014–21

14. Marcellus Shale Coalition 2012 (retrieved 11 October 2012 from http://marcelluscoalition.org/2012/08/american-natural-gas-a-source-of-sustained-economic-growth/)

15. McKenzie L M, Witter R Z, Newman L S and Adgate J L 2012 Human health risk assessment of air emissions from development of unconventional natural gas resources Sci. Total Environ. 424 79–87

16. Mooney C 2011 The truth about fracking Sci. Am. 305 80–5

17. Muller N Z and Mendelsohn R 2007 Measuring the damages of air pollution in the United States J. Environ. Econom. Manag. 54 1–14

18. Muller N Z and Mendelsohn R 2012 Efficient pollution regulation: getting the prices right: corrigendum (mortality rate update) Am. Econ. Rev. 102 613–6

19. NRC (National Research Council) 2010 Hidden Costs of Energy: Unpriced Consequences of Energy Production and Use (Washington, DC: National Academies Press) (www.nap.edu/catalog.php?record_id=12794)

20. NYS DEC (New York State Department of Environmental Conservation) 2011 Revised Draft Supplemental Generic Environmental Impact Statement (EIS) on the Oil, Gas and Solution Mining Regulatory Program (www.dec.ny.gov/data/dmn/rds-geisfull0911.pdf)

21. PA DEP 2011 (retrieved July 2012 from www.paoilandgasreporting.state.pa.us/publicreports/Modules/DataExports/DataExports.aspx)

22. PennEnvironment Research and Policy Center 2012 The Costs of Fracking: The Price Tag of Dirty Drilling's Environmental Damage (www.pennenvironment.org/sites/environment/files/reports/The%20Costs%20of%20Fracking%20vPA_0.pdf)

23. Pennsylvania Department of Environmental Protection 2011 (retrieved 8 July 2012 from www.elibrary.dep.state.pa.us/dsweb/Get/Document-86312/2700-FS-DEP4354.pdf)

24. Pétron G, Frost G, Miller B R, Hirsch A I, Montzka S A, Karion A, Trainer M, Sweeney C, Andrews A E and Miller L 2012 Hydrocarbon emissions characterization in the Colorado Front Range: a pilot study J. Geophys. Res. 117 D04304

25. Ramamurthy K 2012 Personal Communication on Air Emission Inventory Submissions to PA DEP

26. Shogren E 2011 (National Public Radio) (retrieved September 2012 from www.npr.org/2011/06/21/137197991/air-quality-concerns-threaten-natural-gas-image)

27. Skone T, Littlefield J, Eckard R, Cooney G and Marriott J 2012 Role of Alternative Energy Sources: Natural Gas Technology Assessment (NETL/DOE-2012/1539) (www.netl.doe.gov/energy-analyses/refshelf/PubDetails.aspx?Action=View&PubId=435)

28. Steinzor N, Subra W and Sumi L 2012 Gas patch roulette: how shale gas development risks public health in Pennsylvania Earthworks Oil and Gas Accountability Project (www.earthworksaction.org/files/publications/Health-Report-Full-FINAL-sm.pdf)

29. US DOE Argonne National Labs (ANL) 2012 The Greenhouse Gases, Regulated Emissions, and Energy Use in Transportation (GREET) Model, GREET 2 2012 (http://greet.es.anl.gov/)

30. US DOE EIA 2012a Annual Energy Outlook 2012 (Washington, DC: US DOE)

31. US DOE EIA 2012b Monthly Natural Gas Gross Production Report (data for September 2012, www.eia.gov/oil_gas/natural_gas/data_publications/eia914/eia914.html, retrieved November 2012)

32. US EPA 1999 The Benefits and Costs of the Clean Air Act 1990 to 2010 (EPA Report to Congr. EPA-410-R-99-001) (Washington, DC: Office of Air and Radiation, Office of Policy, US Environmental Protection Agency) (www.epa.gov/oar/sect812/1990-2010/chap1130.pdf)

33. US EPA 2008 National Emissions Inventory (NEI) Database (from www.epa.gov/ttnchie1/net/2008inventory.html)

34. US EPA 2012 Air Rules for the Oil and Natural Gas Industry (retrieved July 2012 from www.epa.gov/airquality/oilandgas/actions.html)

35. US GAO 2012 Information on Shale Resources, Development, and Environmental and Public Health Risks (www.gao.gov/assets/650/647791.pdf)

36. Venkatesh A, Jaramillo P, Griffin W M and Matthews H S 2011 Uncertainty in life cycle greenhouse gas emissions from united states natural gas end-uses and its effects on policy Environ. Sci. Technol. 45 8182–9
37. Weber C L and Clavin C 2012 Life cycle carbon footprint of shale gas: review of evidence and implications Environ. Sci. Technol. 46 5688–95

There are several supplemental files that are not available in this version of the article. To view this additional information, please use the citation on the first page of this chapter.

PART IV

URBAN AIR POLLUTION

CHAPTER 9

Modelling Human Exposure to Air Pollutants in an Urban Area

MARCO SCHIAVON, GIANLUCA ANTONACCI,
ELENA CRISTINA RADA, MARCO RAGAZZI, AND DINO ZARDI

Road traffic is one of the main sources of atmospheric pollution in urban areas, where the contribution of traffic flows, congestions and high population density typically increases the exposure of the settled population to pollutants [1-3].

One of the most important family of traffic-related pollutants is represented by nitrogen oxides (NOx), whose main contributors to their emissions in Europe are the mobile sources (38.4%), followed by the energy production sector (21.1%), the commercial, institutional and household sectors (14.8%), energy use in industry (13.4%) and other minor sectors (12.3%) [4].

Unlike nitrogen monoxide (NO), nitrogen dioxide (NO_2) produces adverse effects on human health, such as the development of bronchitis, pneumonia, asthma and the pulmonary growth reduction in chronically ex-

Modelling Human Exposure to Air Pollutants in an Urban Area. Schiavon M, Antonacci G, Rada EC, Ragazzi M, and Zardi D. Revista de Chimie **65**,*1 (2014), http://revistadechimie.ro/pdf/SCHIA-VON%20M.pdf%201%2014.pdf. Reprinted with permission from the authors and the publisher.*

posed adults and children [5, 6]. To preserve the human health, the World Health Organization (WHO) established a limit value of 40μg m^{-3} for the annual mean concentration of NO$_2$. An hourly threshold of 200 μg m^{-3} was also set in order to prevent the effects of acute exposure, such as medium airways inflammations and amplification of allergic reactions [5].

Since emitted NO tends to be rapidly converted to NO$_2$, also within the O$_3$ chemical reactions forced by solar radiation [5], NO$_2$ is commonly found near the emission sources. Thus, NO$_2$ can be considered a typical indicator of air pollution from road traffic in urban areas. Critical situations for European urban areas (such as intense road traffic and particular meteorological conditions due to the complex urban structures) make the compliance of the imposed limits a difficult task. The compactness of urban fabrics, the presence of streets with highly developed buildings in the vertical dimension and the lack of open spaces favour the stagnation of pollutants and increase the population exposure. The structuring elements of the urban fabric are the so-called street canyons, intended as narrow roads laterally delimited by two continuous rows of buildings. At this spatial scale, the street canyon represents the basic geometric unit of the urban fabric and defines the vertical dimension of the urban canopy layer, which is the atmospheric layer included between the soil and the roofs [7], after which the urban boundary layer starts to develop. The canyon length (L), its width (W) and the height (H) of the buildings are relevant parameters for the dispersion of pollutants [8].

The understanding of air quality processes in an urban area is facilitated by high resolution simulations. Numerical models are very important for the dual purpose of simulating air quality scenarios and obtaining fields of concentration in areas not covered by the air quality monitoring stations.

This study aims at presenting a methodology to study the role of urban street canyons in the stagnation of pollutants and to detect critical situations of exposure to air pollutants in a densely built area. This work was carried out by applying the COPERT emission algorithm and the AUSTAL2000 dispersion model to an urban area of the town of Verona (Italy). The choice of road traffic as the only emission source is due to the fact that other sources (e.g., domestic heating and industrial activities) generally emit above the urban canopy and their contribution is not dominant inside canyons with high traffic. The analysis considers NO$_2$ as the reference pollutant to evaluate the contribution of road traffic to human exposure in street canyons.

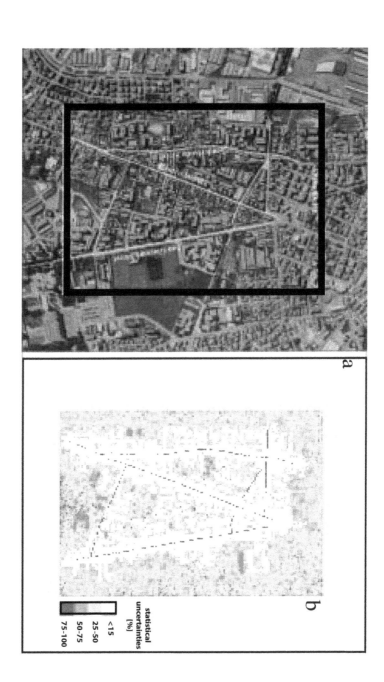

FIGURE 1: a) Computational domain with location of the anemometer, digitization of the streets and the buildings and b) map of the statistical uncertainties for the simulations performed

9.1 EXPERIMENTAL PART

9.1.1 MATERIALS AND METHODS

The town of Verona is located in the Po Plain and is characterized by a continental climate with frequent episodes of thermal inversion [9], especially during the wintertime, like most of the other cities in the Po Plain. Verona is also the junction of important roadways (4 state roads, 1 ring road and 2 highways). From the beginning of 2009, the Municipality has activated a monitoring network of the vehicle fluxes by means of inductive loops, to obtain information about the traffic intensity in selected areas of the town.

As a first step, a preliminary GIS analysis was performed in order to select a particularly critical area of Verona in terms of population density, presence of sensitive receptors (e.g., school buildings) and lack of green areas.

The reconstruction of the vehicle fluxes moved from the data acquired by the traffic monitoring network, consisting in hourly mean fluxes related to a summer and a winter week.

The emissions were calculated by means of the COPERT algorithm, proposed by the European Environmental Agency (EEA) as a tool for evaluating emissions from road traffic within the CORINAIR Programme. The model allows estimating the emissions of the main pollutants related to road traffic: NOx, CO, PM, VOC, CH_4, CO_2, SO_2, Pb and other metals. Information about the vehicle fleet was provided by the Automobile Club d'Italia (ACI) [10]. The calculation of NOx emissions was performed for a whole solar year, in order to compare the concentration maps of the dispersion model with the limit values on annual basis. The hourly vehicle fluxes (together with information on the length of each road) were used to calculate the total NOx emissions (expressed in g s^{-1}) along each street during the whole year, on the basis of each hourly flux, the composition of the vehicle fleet, each street length and the emission factors for each vehicle class. The latter were calculated by assuming an annual mean temperature of 15°C and a mean speed of 45 km h^{-1}, since the streets considered allow a flowing stream of vehicles.

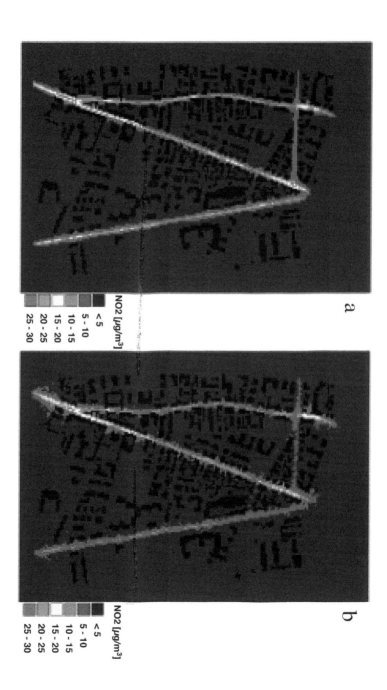

FIGURE 2: a) Annual mean and b) maximal hourly concentrations of NO₂ considering the whole vehicle fleet

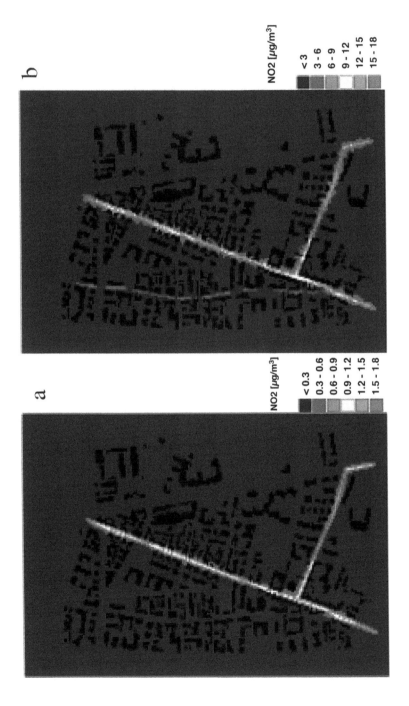

FIGURE 3: a) Annual mean and b) maximal hourly concentrations of NO_2 considering the only contribution of urban buses

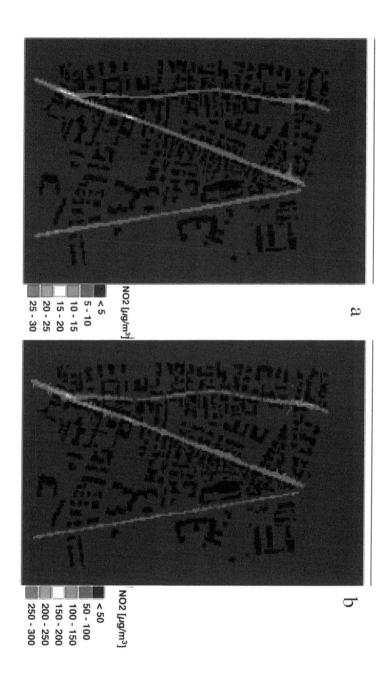

FIGURE 4: a) Annual mean and b) maximal hourly concentrations of NO_2 considering only EURO 2 and later vehicles

After calculating the NOx emissions, the input files for the dispersion model were prepared. The model adopted was AUSTAL2000, a quasi non-stationary three-dimensional lagrangian model, which works on an average of stationary states, opportunely scaled along the flow field. AUST-AL2000 is able to compute the transport of pollutants at a local scale. The meteorological pre-processor (TalDIA) incorporates its own algorithm to assess the effects of buildings on the wind flow, which is quite useful in urban areas, and provides 3D flow fields for the dispersion model. When explicitly defining the obstacles, as in the present case, the parameterized roughness (z_0) adopted used in dispersion algorithms is not used. TalDIA is a diagnostic flow model providing an economic calculation capability based upon profiles of wind and atmospheric stability (according to the Klug-Manier parameterization, conceptually similar to the Pasquill stability). The required data about atmospheric stability classes, wind speed and wind direction for the reference year, were evaluated from a meteorological station managed by the Regional Environmental Protection Agency of Veneto near the study area. In addition, a map of the building heights was elaborated with Quantum GIS and Grass GIS and entered into the model.

Since the emissions calculated by COPERT referred to NOx, as sum of the volumetric fractions of NO and NO_2, it was decided to convert them to NO_2 emissions. For the present case, based on the results from other studies concerning street canyons [11, 12], a constant value of 0.3 was assumed for the NO_2 to NOx concentration ratio. Actually, an empirical relation to calculate the mean annual NO_2 concentration from the corresponding NOx concentration, calibrated on field measurements, was proposed in a previous study [13]. This relation describes well the conversion of NOx in the free field, but underestimates the NO_2 concentration within street canyons, since this approach is not able to consider the effect of stagnation of NO_2 in confined spaces [12].

The emission sources were made to coincide with the streets and parameterized by linear sources. The source heights were set at 0.5 m above the ground, while the vertical extension was assumed to be dependent on the typology of the streets: for street canyons, a vertical extension of 1 m was assumed, whilst, for the remaining streets, the vertical extension was set to 0.5 m. This choice reflects the fact that the stronger mechanical by-produced turbulence within a street canyon from vehicle motions affects the dispersion of pollutants at the source level and contributes to a better mixing of these compounds. A value of 0.9 m was adopted for the surface roughness outside the area covered by the buildings, according to the indications of the CORINE maps, as a result of averaging out the suggested values for terrains with continuous coverage of buildings and terrains with commercial and manufacturing activities [14], as in the case of the area object of this study. The result of the dispersion calculation is the concentration field of the pollutants (in this case NO_2) at 1.5 m above the ground, averaged over subsequent time intervals.

The final aim of this study is the creation of a qualitative exposure map, which can give indications on the most interesting points (with exclusive respect to the domain considered) where to set a monitoring campaign. Exposure depends both on the concentration of the pollutant of interest and on the population density within the study area [15]. The population exposure can be assessed, in a GIS environment, by creating a raster map, based on the product between the concentrations of pollutant and the population density. Thus, the map containing the annual mean concentrations was multiplied by the map of the population density.

9.2 RESULTS AND DISCUSSIONS

The calculation domain is an area of 720 x 1035 m^2 including two street canyons: Centro Street and Scuderlando Street (fig. 1a). The AUSTAL2000 output files are maps of the annual mean and maximal hourly concentrations. In addition, the model creates a map with the statistical uncertainties of the calculated concentrations: precisely, AUSTAL2000 calculates an estimation of the uncertainty, which is directly proportional to the statistical significance of the lagrangian scheme. This map contains, for each cell, the ratio between the standard deviation and the calculated concentration. In this case, satisfying statistical uncertainties were achieved (fig. 1b).

The annual mean concentrations are everywhere lower than the limit value of 40 µg m^{-3}, with a maximum of 27 µg m^{-3} near the junction between the two canyons (fig. 2a). However, only traffic-related NO_2 was taken into account, since the first objective of this work was the estimation of the effect of street canyons in limiting the dispersion of the pollutants.

The second highest value (26 µg m^{-3}) occurs inside Centro Street. Indeed, next to this point, the width of the street canyon is minimal (10 m).

Similar considerations can be expressed about the map of the maximal hourly concentration: these reaches the highest levels at the same points of the annual mean concentration map (fig. 2b). The maximal concentration (270 µg m^{-3}) occurs next to the junction between Centro Street and Scuderlando Street, while a concentration of 227 µg m^{-3} occurs in the narrowest stretch of Centro Street. Lower concentrations occur within Scuderlando Street, although a value of 200 µg m^{-3} is observed in the northern stretch of the canyon, due to a H to W ratio higher than 1 and to intense traffic fluxes. It is interesting to highlight the fact that the concentrations along San Giacomo Street, a road without characteristics of street canyon but with the highest traffic fluxes, result everywhere lower than the concentrations obtained within the two streets canyons.

The contribution of the public transportation (urban buses) was also studied. As shown in figure 3a and 3b, its contribution to the NO$_2$ concentration is minimal. The maximal hourly contribution to NO$_2$ concentrations occurs in Centro Street (18 µg m^{-3}). At the same point, the maximal hourly concentration calculated for the whole car fleet was 210 µg m^{-3}. The highest annual mean concentration contribution is 2 µg m^{-3}, which also occurs at the same point. The same cell showed a concentration of 26 µg m^{-3} for the simulation with the whole fleet. Therefore, the contribution of urban buses to the total NO$_2$ concentrations is about 8%.

To assess the effectiveness of possible measures to restrict the circulation to the most recent (and, thus, less polluting) vehicles, a new simulation was performed with the exclusion of EURO 0 and EURO 1 vehicles. Both the concentration maps show a sensible reduction of NO$_2$ concentrations with respect to the simulation with the entire vehicle fleet (fig. 4a and 4b). Similarly to the previous simulations, the highest annual mean concentration (17 µg m^{-3}) occurs near the junction between Scuderlando Street and Centro Street; at the same point, the maximum hourly concentration also occurs (168 µg m^{-3}). Since the highest concentrations of the previous simulation were 27 and 270 µg m^{-3} (respectively for the annual mean and the maximal hourly concentration), in both of the cases a restriction to the circulation of EURO 0 and EURO 1 vehicles (and a following reduction of the fluxes) leads to a concentration reduction by about 40%. This result is in line with the findings of a study that examined the effects of traffic restrictions on the NO_2 concentrations within an urban area [16].

As a final step, the level of exposure of the population inside the street canyons was estimated (fig. 5), in accordance with the methodology presented in the previous section. It is important to say that this map provides that zones with the highest exposure levels are located near the intersection between the two canyons and at the northern stretch of Centro Street.

9.3 CONCLUSIONS

The results of the simulations highlight the problem of the dispersion of pollutants within a complex urban area; in particular, the simulations pointed out to what extent the morphology and density of the buildings are important to favour the stagnation of pollutants within the urban fabric. This explains why the highest NO_2 concentrations occur within the two identified street canyons, although the traffic fluxes were lower than those observed for other roads without buildings. The simulations allowed to identify the most critical zones, thus providing a useful insight for the planning of monitoring campaign of air quality (e.g., with the use of passive samplers, as well as mobile or even fixed stations). Simulations on shorter periods would allow detecting the most critical periods of the year and, then, would give indications about the most interesting period for the samplings.

As a future step, the combination of all the most relevant NOx emission sources (i.e., traffic, domestic heating and industrial processes) would provide a tool to support decisions for the urban planning. This would be especially important for the location of sensitive activities such as hospitals or schools.

The concept of exposure is of great interest, especially due to the fact that the current legislation does not consider the location of the emission sources and the settled population. Therefore, exposure maps provide an interesting starting point to conduct zoning analyses and to detect the areas where the population is more exposed to potential health risks.

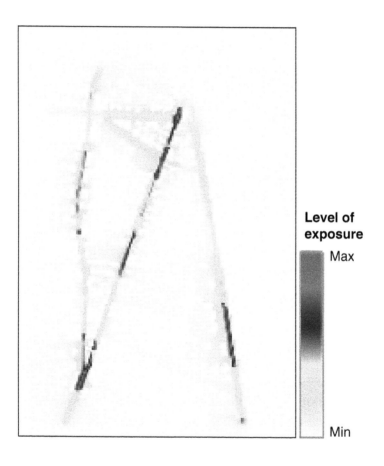

FIGURE 5: Qualitative map of the population exposure to NO_2 emitted by road traffic inside the urban area under investigation

REFERENCES

1. Torretta, V., Rada, E.C., Panaitescu, V.N., Apostol T., Sci Bull., 74, N. 4, Serie D, 2012, P. 141.
2. Ionescu, G., Apostol, T., Rada, E.C., Ragazzi, M., Torretta, V. Sci Bull., 75, N. 2, Serie D, 2013, P. 175.
3. Rada, E.cC, Ragazzi, M., Brini, M., Marmo, L., Zambelli, P., Chelodi, M., Ciolli, M. Sci Bull., 74, N. 2, Serie D, 2012, P. 243.
4. EEA – European Environment Agency, Nitrogen Oxide (Nox) Emissions, Http://Www.eea.europa.eu/Data-And-Maps/Indicators/Eea-32-Nitrogen-Oxides-Nox-Emissions-1/Assessment.2010-08-19.0140149032-1, 2012.
5. WHO - World Health Organization, Air Quality Guidelines, Global Update 2005, Particulate Matter, Ozone, Nitrogen Dioxide And Sulphur Dioxide, Druckpartner Moser, Germany, 2006.
6. Kulkarni, N. And Gridd, L., J. Paediatr. Child Health, 18, 2008, P. 238.
7. Sini, J.F., Anquetin, S., Mestayer, P.G., Atmos. Environ., 30, 1996, P. 2659.
8. Oke, T.R., Energy Build., 11, 1988, P. 103.
9. Andrighetti, M., Zardi, D., De Franceschi, M., Meteor. Atmos. Phys., 103, 2009, P. 267.
10. ACI – Automobile Club D'italia, Http://Www.aci.it/Laci/Studi-Ericerche/Dati-E-Statistiche/Autoritratto.html, 2010.
11. Vardoulakis, S., Valiantis, M., Milner, J., Apsimon, H., Atmos. Environ., 41, 2007, P. 4622.
12. Düring, I., Bächlin, W., Ketzel, M., Baum, A., Friedrich, U., Wurzler, S., Meteorol. Z., 20, 2011, P. 67.
13. Romberg, E., Bösinger, R., Lohmeyer, A., Ruhnke, R., Röth, E., Gefahrst.-Reinhalt. Luft, 56, 1996, P. 215.
14. Silva, J., Ribeiro, C., Guedes, R., Roughness Length Classification Of Corine Land Cover Classes, Technical Report, Megajoule-Consulting, Mona Vale, Nsw, Australia, 2007.
15. Heimann, D., Clemente, M., Elampe, E., Only, X., Miège, B., Defrance, J., Baulac, M., Suppan, P., Schäfer, K., Emeis, S., Forkel, R., Trini Castelli, S., Anfossi, D., Belfiore, G., Lercher, P., Rüdisser, J., Uhrner, U., Öttl, D., Rexeis, M., De Franceschi, M., Zardi, Cocarta, D., Ragazzi, M., Antonacci, G., Cemin, A., Seibert, P., Schicker, I., Krüger, B., Obleitner, F., Vergeiner, J., Grießer, E., Botteldooren, D., Renterghem, T. Van, "Air Pollution, Traffic Noise And Related Health Effects In The Alpine Space", University Of Trento, Dica, 2007, P. 335.
16. Oduyemi, K.o.k., Davidson, B., Sci. Tot. Environ., 218, 1998, P. 59

CHAPTER 10

New Considerations for PM, Black Carbon, and Particle Number Concentration for Air Quality Monitoring Across Different European Cities

C. RECHE, X. QUEROL, A. ALASTUEY, M. VIANA, J. PEY,
T. MORENO, S. RODRÍGUEZ, Y. GONZÁLEZ,
R. FERNÁNDEZ-CAMACHO, A. M. SÁNCHEZ DE LA CAMPA,
J. DE LA ROSA, M. DALL'OSTO, A. S. H. PRÉVÔT, C. HUEGLIN,
R. M. HARRISON, AND P. QUINCEY

10.1 INTRODUCTION

Health impact of ultrafine particles (UFP) has motivated a great deal of ambient aerosol research in recent years. Several studies suggest that UFP disproportionately induce oxidative stress in cells and are more toxic compared to larger particles of similar composition (Li et al., 2003; Nel et al.,

New Considerations for PM, Black Carbon, and Particle Number Concentration for Air Quality Monitoring Across Different European Cities. © *Reche C et al.* Atmospheric Chemistry and Physics *11 (2011). doi:10.5194/acp-11-6207-2011. Licensed under Creative Commons Attribution 3.0 Unported License, http://creativecommons.org/licenses/by/3.0/.*

2005). Research studies have indicated that fine particles may be more toxic because a large proportion of these particles are derived from traffic-related, industrial, and domestic emissions which contain abundant transition metals (Anderson et al., 2001; Klemm et al., 2000; Schwartz et al., 2002; von Klot et al., 2002). Furthermore, UFP have also been suggested to be more toxic because of the large surface area available for biologic interactions with lung cells (Chio et al., 2008). Epidemiological studies (Akinson et al., 2010; Stolzel et al., 2007) have shown a clear association of urban nanoparticle exposures with adverse cardiovascular health outcomes.

Because of the low influence of UFP on PM mass concentration (the current metric used in European air quality legislation), the number concentration (N) can be a better descriptor of the variability of UFP. In fact, the combination of number and size distribution of N may elucidate on the primary or secondary production of these UFP.

In many large cities of Europe standard air quality limit values of PM are exceeded. Emissions from road traffic and biomass burning are frequently reported to be the major causes of such exceedances (EEA, 2010). As a consequence of these exceedances a large number of air quality plans, most of them focusing on traffic emissions, have been implemented in the last decade. In spite of this implementation, a number of cities did not record a decrease of PM levels. Thus, the question remains: is the efficiency of air quality plans overestimated? Do the road traffic emissions contribute less than expected to ambient air PM levels in urban areas? Or do we need a more specific metric to evaluate the impact of the above emissions on the levels of urban aerosols?

A number of studies have reported a strong association between BC and road traffic (Hamilton and Mansfield, 1991; Watson et al., 1994; Pakkanen et al., 2000) and biomass burning (Sandradewi et al., 2008) emissions. While BC aerosols are not the only cause of adverse health effects due to particles, they are a major factor, specially the ultrafine BC. Indeed, the recent WHO report concludes that "combustion-derived aerosols are particularly significant in terms of their health effects" (WHO, 2003).

On the other hand, number concentrations in urban areas are also highly influenced by primary vehicle exhaust emissions (Morawska et al, 2002; Bukowiecki et al., 2003; Hueglin et al., 2006; Rodriguez and Cuevas, 2007, Perez et al., 2010). These emissions show bimodal size distribution, with a nucleation mode below 30 nm and a carbonaceous mode peaking between 50–130 nm (Morawska et al., 1998, Casati et al., 2007). Therefore, a number of studies pointed out that exposure to road traffic emissions may be properly evaluated by combining ambient air measurements of Black Carbon (BC) with N concentrations (Fischer et al., 2000; Harrison et al., 2004; Janssen et al., 1997; Smargiassi et al., 2005; Rodriguez and Cuevas, 2007), since nowadays these parameters seem not to be properly controlled by air quality limit values.

Although most of UFP in urban atmospheres are related to vehicle exhaust emissions, its origin may be both primary and secondary (Wehner et al., 2002; Dunn et al., 2004; Van Dingenen et al., 2004). Gaseous pollutants from vehicle exhaust may yield a high aerosol production depending on the ambient air conditions. For example, Casati et al. (2007) observed that low ambient temperature and high relative humidity favour secondary formation processes, while Shi and Harrison (1999) associated these processes with high dilution after the emission. Furthermore, Wehner et al. (2009) reported evidence on the fact that under high engine load conditions a favourable setting for secondary particle formation is given within short distance of the exhaust plume. Such conditions (e.g., acceleration) are typical of urban driving. These particles in the vehicle exhaust mainly comprise organic compounds and sulphuric acid and are frequently one of the most important fractions in terms of number concentration (Kittelson et al., 1998).

New secondary particle formation in ambient air is mostly attributed to nucleation and cluster/particle growth by condensation of photo-oxidised vapours (Morawska et al., 2008; Dunn et al., 2004) occurring some time after the emission (hours to days). The outcome of UFP depends widely on the pollutants concentrations in the air, thus when the urban atmosphere is highly polluted, the semi-volatile species condense onto pre-existing particles (Wichmann et al., 2000; Zhang et al., 2004; Imhof et al., 2006); however, when low PM pollution levels occur, the semi-volatile species may result in large numbers of nucleation-derived aerosols (Hameri et al., 1996; Ronkko et al., 2006). As previously reported, urban areas with high solar radiation intensities are favourable scenarios for nucleation processes (Johnson et al., 2005; Moore et al., 2007; Pey et al., 2008, 2009; Fernandez-Camacho et al., 2010; Cheung et al., 2010). Elevated solar radiation intensities not only provide enough energy for gaseous precursors to nucleate, but favour the dilution processes as a result of the growing of the mixing layer and the activation of mountain and sea breezes. In case of coastal cities, such as Barcelona and Santa Cruz de Tenerife, the significant SO_2 emissions from shipping may be transported towards the city with the sea breeze, enhancing the nucleation processes. The combination of the oxidation products of SO_2 and VOCs are ideal for efficient nucleation (Metzger et al., 2010).

Once created, nucleated particles may undergo different processes in the atmosphere (Boy and Kulmala, 2002; Qian et al., 2007; Park et al., 2009; Gao et al., 2009; Cheung et al., 2010): (1) burst of nucleation particles without subsequent growth into larger particles, and (2) nucleation coupled with growth. Factors governing the evolution of nucleated particles without subsequent growth are not well-documented (Johnson et al., 2005; Moore et al., 2007; Pey et al., 2008; Park et al., 2009; Gao et al, 2009) and further research is needed to elucidate the occurrence of this process in urban environments. At coastal urban sites, the development of sea breezes (enriched in SOx) simultaneously with the highest solar radiation could be probably related with this process. Both factors favor nucleation processes, but the significant increase in wind speed can be responsible for a high dilution of the condensable gases involved in growth. Thus, studies have confirmed that the growth rate depends on temperature and concentration of available condensable vapors (Kulmala et al., 2004). In some cases it has been suggested that H_2SO_4 condensation typically accounts for about 10–30 % of the observed growth (Weber et al., 1997; Boy et al., 2005), whereas VOCs account for more than 70 % of the material for the particle growth.

The daily cycle of N evidences significant differences when comparing cities with distinct meteorological conditions. Daily N cycles showing new particle formation coinciding with sea breezes blowing inland have recently been observed in the urban background of coastal cities in southwestern Europe, such as Barcelona, Santa Cruz de Tenerife and Huelva (Pey et al., 2008; Rodrıguez et al., 2008; Fernandez-Camacho et al., 2010; Perez et al., 2010), as well as in Brisbane, Australia (Mejia et al., 2009; Cheung et al., 2010). In other cases such as Beijing, daily formation of nucleation mode particles was coincident with the arrival of cleaner air masses (Zong-bo et al., 2007).

FIGURE 1: Location of the six cities selected in the present study.

The increase in knowledge of UFP has not been accompanied by more in-depth research about the main factors governing differences in the secondary formation of particles between urban sites under a variety of emission sources and climate/geographic conditions. In this context, the main goal of this paper is to study the causes responsible for the variability of levels of N, BC, PM and gaseous pollutants at a selection of air quality monitoring sites representative of different climate zones and urban environments in Europe, with especial focus on the process of formation of secondary UFP with high influence on the variability of N.

10.2 MONITORING SITES

Seven monitoring sites situated in six major European cities (Fig. 1) with different climatic and meteorological patterns were selected for the study. The selection of the cities/sites intended to cover different climatic zones across Europe (central, western and southern Europe), as well as urban environments (urban background and traffic, urbanindustrial and urban-shipping-influenced sites). The following sites also cover a relatively wide range of road traffic density, urban architecture and fleet composition (Table 2):

- A southern-European urban background site (Barcelona, BCN) characterized by a very dense road traffic network, but also by the influence of industrial and shipping emissions.
- A north-European urban background site (North Kensington, UK, NK) in the grounds of a school in a residential area 7 km to the west of central London.
- Two urban traffic sites situated directly at the kerbside of very busy roads: Bern (Switzerland, central-Europe), with a traffic density of 25 000 vehicles/day, and Marylebone road in London (UK, northern-Europe), with a traffic density of 80 000 vehicles/day.
- A central-European urban background site (Lugano, Switzerland, LUG) situated in a park in the south of the Alps.
- A southern-European urban background site (Huelva, HU) highly influenced by the emissions from a large industrial estate including copper metallurgy, petrochemical and fertilizing plants.
- A subtropical island urban background site (Santa Cruz de Tenerife, Spain, SCO) on the western side of a 4-lane road running along the shore, with

influence of the emissions from a nearby large harbour and an industrial-petrochemical estate.

10.2.1 BARCELONA, BCN (URBAN BACKGROUND SITE IN A CITY WITH DENSE TRAFFIC)

Barcelona is located in north eastern Spain (41° 230 05° N; 02° 070 09° E; 68 m a.s.l.), in the western Mediterranean Basin. It is the tenth most populous city in Europe, with about 1.7 million inhabitants. The urban architecture and dynamism around Barcelona account for the highest road traffic density of Europe (6100 cars km^{-2}, much more than in most European cities with 1000–1500 cars km^{-2}). As shown in Table 1, Barcelona is characterized by a high proportion of diesel cars, motorbikes, heavy duty vehicles, and a large proportion of the use of private cars for the daily mobility. Furthermore, Barcelona has one of the main harbours in the Mediterranean Basin, with the highest number of cruise ships for tourists in Spain, being a significant focus of emissions of atmospheric pollutants, which are very often transported across the city by sea breeze. Finally, a wide range of intensive industrial activities, three natural gas power stations and two city waste incinerators are also based in the metropolitan area.

In addition to the local PM emissions, Saharan dust outbreaks reach the Barcelona area in the order of 7–10 events per year, with a major frequency in the summer and winter–spring periods (Rodrıguez et al., 2001).

The transport and dispersion of atmospheric pollutants within BCN are controlled mainly by fluctuating coastal winds which typically blow in from the sea during the day and, less strongly, from the land during the night. The sea breezes (originating from the 120–180° sector) are at their strongest around midday, when the boundary layer height maximizes (Perez et al., 2004). The average annual solar radiation is 180 W m^{-2} and at midday values range between 400 and 950 W m^{-2} within a whole year. The annual accumulated precipitation is 500 mm.

Measurements were carried out at an urban background monitoring site located in southwest Barcelona, being influenced by vehicular emissions from one of the city's main traffic avenues, located at approximately a distance of 300 m with a mean traffic density of 132 000 vehicles/day. Thus, the site is an urban background one, but it is located in a city with very high road traffic, and influenced by the emissions of one of the largest arterial roads of the city.

10.2.2 LONDON (URBAN BACKGROUND AND TRAFFIC SITES)

London has a population of 7.6 million, whilst the Greater London metropolitan area has between 12.3 and 13.9 million, making it the largest in the European Union (Wikipedia, 2010). The traffic density in London in 2009 was 1317 registered vehicles km^{-2} of which 1134 km^{-2} were cars. As shown in Table 1, London is characterized by a lower proportion of diesel cars, motorbikes, heavy duty vehicles, and also low proportion of the use of private cars for the daily mobility.

On an annual basis, mean solar radiation is 70 W m^{-2}, with values ranging from 5 to 760 W m^{-2} at midday. The daily pattern of the boundary layer height shows maximum from 12:00 to 15:00 UTC, contributing to decrease atmospheric pollutants concentration.

For the present study two air quality monitoring sites were selected:

10.2.2.1 LONDON, MARYLEBONE, MR (URBAN TRAFFIC SITE)

The London–MR monitoring site is located on the kerbside of a major arterial route in London that is heavily trafficked (51∘310 96∘ N; 00∘9 0 5500W; 27 m a.s.l.). The surrounding area is a street canyon frequented by pedestrians because of tourist attractions and shops. High PM_{10} concentrations are measured at MR and the permitted number of days with concentrations above the limit value was exceeded in 2005. This site is classified as a roadside site. The MR supersite belongs to the London Air Quality Network. The surrounding buildings form an asymmetric street canyon (height-to-width ratio of about 0.8). Traffic flows of over 80 000 vehicles/day pass the site on six lanes with frequent congestion. Braking is frequent near the measurement site due to the presence of traffic lights 50 m to the west and an intersection to the east. The instruments are in a cabin and sampling inlets are less than 5 m from the road. Local PM_{10} emissions are strongly dominated by the heavy-duty vehicles that represent less than 10% of the traffic (Charron and Harrison, 2005).

10.2.2.2 LONDON, NORTH KENSINGTON, NK (URBAN BACKGROUND)

This is sited in the grounds of Sion Manning School in St Charles Square, North Kensington (51° 310 1600 N; 00° 1204800 W; 27 m a.s.l.), surrounded by a mainly residential area. The NK site is located about 4 km to the west of MR site.

TABLE 1: Features of fleets and commuting of the 6 selected cities.

	Barcelona	London	Bern	Lugano	Huelva	Sta. Cruz
% Diesel fleet	45[a]	?	17[d]	17[d]	55[a]	24[a]
% New cars diesel	70[a]	33[c]	30[d]	27[d]	?	?
Commuting						
Private car	40[b]	37[c]	35[e]	48[f]	45[g]	?
Public T.	28[b]	41[c]	18[e]	12[f]	1[g]	?
Pedestrian	32[b]	21[c]	47[e]	40[f]	52[g]	?
Fleet						
Motorbike	29[b]	4[c]	18[e]	16[f]	7[g]	6[g]
Passenger and LDV	63[b]	93[c]	77[e]	81[f]	87[g]	87[g]
Bus	2[b]	1[c]	1[e]	1[f]	0.2[g]	0.4[g]
HDV	3[b]	1[c]	1[e]	1[f]	3[g]	4[g]
Other	2[b]	1[c]	3[e]	2[f]	1[g]	2[g]

[a] *Direccion General de Trafico, Spain: http://apl.dgt.es/IEST2.* [b] *Baldasano et al. (2007).* [c] *Department of Transport, UK: http://www.dft.gov.uk.* [d] *Swiss Federal Statistical Office: http://www.bfs.admin.ch.* [e] *Ecoplan (2007) Auswertung Mikrozensus 2005 fur den Kanton Bern, Report, Bern: "http://www.bve.be.ch/bve/de/index/mobilitaet/mobilitaet verkehr/ mobilitaet/grundlagen mobilitaet.html.* [f] *www.tiresia.ch.* [g] *Oral communications from local councils.*

10.2.3 BERN, BERN (URBAN TRAFFIC SITE)

Bern is a city with a population of 0.125 million and the fourth most populous city in Switzerland. The urban area of Bern including neighbouring communities has a population of 0.35 million. Bern is the capital of Switzerland and the Canton of Bern. The city is located north of the Alps in the Swiss plateau (46° 570 0300 N, 07° 260 2700 E, 536 m a.s.l.). Attending to the typical meteorological feature, the mean annual solar radiation is 130 W m^{-2} and it ranges from 15 W m^{-2} in winter to 840 W m^{-2} in summer at noon. The accumulated precipitation in the city is about 1700 mm per year. The monitoring site is located in a busy street canyon in central Bern (20 000–30 000 vehicles/day). The economy of Bern is dominated by public authorities and small and medium sized enterprises from different sectors. There are, however, no major industries with especially high emissions or air pollutants. As shown in Table 1, Bern is characterized by a low proportion of diesel cars and heavy duty vehicles, high proportion of motorbikes.

10.2.4 LUGANO, LUG (URBAN BACKGROUND SITE)

Lugano is a city with 55 000 inhabitants and a total of 145 000 people living in Lugano and the neighbouring community. The city lies at the edge of Lake Lugano at the southern foothills of the Alps (46°000 4000 N, 08° 570 2600 E, 281 m a.s.l.) and is surrounded by three mountains with elevations of around 1000 m a.s.l. As shown in Table 1, Lugano is characterized by a low proportion of diesel cars and heavy duty vehicles and a high proportion of motorbikes. There are no major important industrial sources of air pollutants within the close vicinity of Lugano. However, Lugano lies close to the Italian border and is strongly influenced by emissions from the Lombardy region.

Precipitations are frequent, with an accumulated value of 3800 mm per year. Mean solar radiation is 150 W m^{-2} on average, at midday this parameter is between 3 and 900 W m^{-2}.

The measurement site in Lugano represents an urban background situation. The site is located on campus of the University of Lugano in the city center. About 50 m to the east is a busy urban road (CorsoElvezia), buildings protect the site from direct road traffic emissions towards all other wind directions.

TABLE 2: Main information about the monitoring sites selected for the study.

	Longitude	Latitude	Altitude (m a.s.l.)	Station type
Barcelona (ES)	02° 070 3300 E	41° 230 5500 N	80	Urban background
Lugano (CH)	08° 570 2600 E	46° 000 4000 N	281	Urban background
North Kensington (London, UK)	00° 120 4800 W	51° 310 1600 N	27	Urban background
Bern (CH)	07° 260 2700 E	46° 570 0400 N	536	Urban traffic
Marylebone Road (London, UK)	00° 090 5500 W	51° 310 9600 N	27	Urban traffic
Huelva (ES)	05° 560 2400 W	03° 150 2100 N	10	Urban industrial
Santa Cruz de Tenerife (ES)	16° 180 3300 W	28° 290 2000 N	52	Urban background

10. 2.5 HUELVA, HU (URBAN BACKGROUND SITE INFLUENCED BY INDUSTRIAL EMISSIONS)

Huelva is located south western Spain (37° 150 0 ° N, 6 ° 5700 ° W, 54 m a.s.l) and has a population of around 0.15 million. In addition to the typical urban emissions (with a high proportion of diesel cars), aerosol precursors are emitted at the south of the city, where two large industrial estates are located: Punta del Sebo and Nuevo Puerto, both near the harbor of Huelva. The Punta del Sebo Industrial Estate includes the second smelter factory in Europe, where SO_2, H_2SO_4, As, Sb, Pb, Zn and Sn emissions are well documented. Phosphoric acid production plants are also installed in this industrial estate. NH_4^+ and Na phosphate, phosphoric acid, sulphuric acid and sodium silicate atmospheric emissions may be expected from these industrial activities. The most important air pollutant emissions in Nuevo Puerto occurs in a petroleum refinery, resulting in emissions of volatile hydrocarbons, SO_2, NOx, NH_3, Ni and V. Shipping emissions occur also in the important industrial harbor of Huelva. In all these cases sea-to-land winds result in the inland transport of aerosols and their precursors (e.g. SO_2) affecting the city of Huelva.

The city of Huelva is also affected by natural PM contributions such as North African dust outbreaks, which produce a generalized increase in mass levels of particulate matter. The annual frequency of this natural phenomenon is calculated to be 19% of days in southern Spain (Sanchez de la Campa et al., 2007). This area is characterized by a dry weather, with a yearly accumulated rainfall of 450 mm. The solar radiation is very elevated, reaching average values of 1200 W m^{-2} on an hourly basis at midday. The dispersion and transport of air pollutants in this area are highly influenced by the topographic settings. At night, the wind mostly blows from the north and during daylight, southern airflows linked to thermally driven breezes predominate. This sea breeze favors the entry of industrial plumes and is associated with an increase in ozone concentrations (Millan et al., 2002). The boundary layer height maximizes from 12:00–15:00 UTC.

Measurements were carried out at an urban background monitoring site placed at the University Campus, on the northeast corner of the city of Huelva, 7 km from Punta del Sebo and 14 km from Nuevo Puerto industrial areas. This monitoring station belongs to the air quality network of the Autonomous Government of Andalusia. The closest roads lie about 500 and 1000 m to the west and the east of the measurement site.

10.2.6 SANTA CRUZ DE TENERIFE, SCO (URBAN BACKGROUND SITE UNDER THE INFLUENCE OF SHIPPING AND INDUSTRIAL EMISSIONS)

Santa Cruz de Tenerife is a city with around 0.223 million population located in the Canary Islands – Spain (28∘290 200 N, 16∘180 3300 W; 52 m a.s.l.). It is located at the bottom of the southern slope of the Anaga ridge and the eastern slope of the NE to SW ridge crossing the Island. This topographic setting protects the city from the trade winds (NNE) that blow over the ocean (Guerra et al., 2004). The main sources of pollutants in the city are: vehicle exhaust emissions (with a low proportion of diesel cars), emissions of ships and cargo operations in the harbour and an oil refinery located in the southern side of the city (Rodr´ıguez and Cuevas, 2007; Rodrıguez et al., 2008).

The urban scale transport of air pollutants in Santa Cruz de Tenerife is mainly driven by breeze circulation. This breeze is characterised by inland (westward) airflows during daylight (3–4 m s^{-1}) and a slight seaward (eastward) airflow at night (1 m s^{-1}). Inland breeze blowing starts at 08:00 UTC and is characterised by an abrupt shift in wind direction (Rodriguez et al., 2008). Solar radiation is 250 W m^{-2} on annual average, with a maximum of 1200 W m^{-2} at midday in summer, coinciding with the maximum height of the mixing layer.

Measurements were performed in the Santa Cruz Observatory. This is a coastal urban background site mainly influenced by vehicle exhaust and harbour emissions, and also by the emissions from a petrochemical estate. In addition to the local emissions, about 54 natural Saharan dust events occur along the year (Alonso-Perez et al., 2007), often resulting in PM$_{10}$ >100 μg m^{-3} (Viana et al., 2002).

10.3 INSTRUMENTATION

Data on the instrumentation deployed at the different sites is summarized in Table 3. For the present study hourly averaged 2009 data were collected for each parameter and monitoring site.

It is relevant to note that differences in BC and N concentrations can be partly derived from the instrumentation. The latter were measured with a total of three models of Condensation Particles Counters (CPC) which counted particles larger than 2.5 nm (TSI 3025 and TSI 3776), 5 nm (TSI 3785) or 7 nm (TSI 3022A). The use of CPCs with different cut sizes is likely to influence the results somewhat, as a largest cut size can imply an underestimation of N, resulting in lowest N/BC ratios.

As regards BC, the Multi-Angle Absorption Photometer (MAAP) (ThermoTM, model Carusso 5012) instrument calculates absorbance from particles deposited on the filter using measurements of both transmittance and reflectance at two different angles. The absorbance is converted to the mass concentration of BC using a fixed mass absorption coefficient at 637 nm (Muller et al., 2010) of 6.6 m^2 g^{-1} recommended by the manufacturer.

Nevertheless, experimental results showed average absorption coefficients of 9.2 m^2 g^{-1} in BCN, 12.1 in LUG, 10.9 in Bern, 10.3 in HU and 9.8 in SCO. Results were obtained by in situ determining elemental carbon (EC) for high volume samples of 24 h by means of the Thermo Optical Transmittance technique (Birch and Cary, 1996) using a Sunset Laboratory OC-EC analyser and the default temperature steps of the EUSAAR2 program (Cavalli et al., 2010). These experimental conversion factors were used in this study.

On the other hand, the Aethalometer (Magee AE 21) measured at 880 nm and a mass absorption cross section of 16.6 m^2 g^{-1} is recommended by the manufacturer to convert the observed light attenuation to the mass concentration of BC. In 2009, high correlation coefficients were obtained between BC mass and EC concentrations in North Kensington ($r^2 = 0.86$) and Marylebone (0.77) and the slope of the regression line was 0.97 and 1.19, respectively. Thus, the calculation of the experimental absorption coefficients, by means of the determination of EC, showed a value higher than the recommended one (16.6 m^2 g^{-1}) in Marylebone (19.7) and closed to it in North Kensington (16.3).

Therefore, in order to perform an accurate comparison between sites, BC concentrations were determined using the experimental absorption coefficients (σ in m^2 g^{-1}) according the equation by Petzold and Schonlinner (2004) (1): ¨

$$BC\ (\mu g\ m^{-3}) = \sigma_{ap}\ (M\ m^{-1})/\sigma\ (m^2 g^{-1}) \tag{1}$$

where σ_{ap} are the absorption coefficient measurements in M m^{-1}.

Although the influence of possible coatings of BC particles by organic materials has not been taken into consideration, the correction of BC values with local EC concentrations carried out in this study favours the comparability between BC in the different cities. This influence could modify the N/BC ratios calculated in this work for the different periods of the day, but it would not alter the trends described.

On the other hand, in spite of this correction, the use of different instrumentations can still affect the seasonal trends of light absorbing carbonaceous aerosols. However, correlations between BC and EC measurements considering the whole year were significant in all the sites under study.

Regarding gaseous pollutants instrumentation, it is important to mention that NO_2 measurements can be overestimated because of interferences of oxidized nitrogen compounds in the conventional instruments equipped with molybdenum converters (Steinbacher et al., 2007).

10.4 RESULTS AND DISCUSSION

10.4.1 LEVELS OF ATMOSPHERIC POLLUTANTS

For the analysis and description of the trends observed for the different parameters, stations have been classified into three groups: (1) traffic stations (MR and Bern), (2) urban background stations (BCN, LUG and NK) and (3) urban background stations with special characteristics (industrial influence in the case of HU and subtropical island conditions, with the influence of shipping and industrial emissions, in the case of SCO). Table 4 lists the 2009 average values of the parameters measured in each station during the sampling period.

Average levels of PM_{10} range from 18 to 32 µg m^{-3}. As expected, the highest values (27–32 µg m^{-3}) were registered for kerbsides and also in the BCN urban background site, due to its proximity to one of the largest arterial roads of the city.

Levels of BC at urban background sites range from 1.7 to 1.9 µg m^{-3} in BCN, LUG and NK. Thus, despite differences in percentages of diesel vehicles between sites, BC levels are almost equal. This fact may be explained by differences in meteorology and by considering the importance of biomass burning emissions in Lugano and coal fired power plants around London, which contribute to increase BC outputs (Szidat et al., 2007; Bigi and Harrison, 2010). Levels of BC were very high in MR, with an annual average of 7.8 µg m^{-3} and decrease down to 3.5 µg m^{-3} in Bern, accordingly with the lower vehicle flow. On the other hand, HU (0.7 µg m^{-3}) and SCO (0.8 µg m^{-3}) recorded the lowest concentrations of BC owing to a lesser impact of traffic in these smaller cities and favourable dispersive conditions (SCO). It is important to note that the ratio PM_{10}/BC is higher at urban background sites (9–18) than at traffic sites (4–8) due to the relative prevalence of secondary compounds. Maximum values of this ratio are observed in HU and SCO (26–33) as a consequence of the important contributions of dust, and also sea salt in SCO.

As regards N, 2009 averages ranged from 12 000–18 000 cm^{-3}, including the industrial site (with the highest levels), with this range increasing to 22 000–28 000 cm^{-3} at traffic sites due to the direct impact of primary exhaust emissions. N/BC ratios range from 6 to 6×10^6 particles ng^{-1} BC, with the exception of MR (2.8×10^6 particles ng^{-1} BC), HU (25×10^6 particles ng^{-1} BC) and SCO (15×10^6 particles ng^{-1} BC). The reasons behind these unusually low and high N/BC ratios will be discussed in the following sections.

Concerning gaseous pollutants, SO$_2$ levels reach between 1.8 µg m^{-3} (LUG) and 3.5 µg m^{-3} (SCO) in the urban background stations. The slightly higher levels measured at BCN and SCO are caused by the significant influence of shipping emissions. Maxima SO$_2$ levels were recorded at the traffic site of MR (6.7 µg m^{-3}), as a direct effect of exhaust emissions (according to data from the UK National Atmospheric Emission Inventory, the emission factor of SO$_2$ from road vehicle engines is 11.5 Kilotonne Mt^{-1} fuel consumed) and at HU (9.2µg m^{-3}), due to the emissions from a large copper smelter, fertilizer and oil refinery industries.

NO concentrations reach 7–16 µg m^{-3} at urban background sites and 27–106 µg m^{-3} at the traffic ones. The lowest values were registered for LUG, HU and SCO. The same trends were observed for CO concentrations, which ranged from 0.2 to 0.4 mg m^{-3} at urban background sites and from 0.5 to 0.7 mg m^{-3} at traffic locations, as this gaseous pollutant reflects proximity and intensity of the traffic flow. The NO/BC ratio ranges from 7 to 10 in most of the sites, with the exception of LUG (5), probably due to a higher influence of biomass burning, increasing BC but not proportionally NO levels, and MR (14), as a result of the proximity to fresh road traffic emissions. The influenced of biomass burning in Lugano is also indicated by the diurnal cycle of CO and PM concentrations, with a more pronounced peak in the evening than the one obtained for NOx levels. Similar results have been reported in previous studies in nearby regions (Sandradewi et al., 2008).

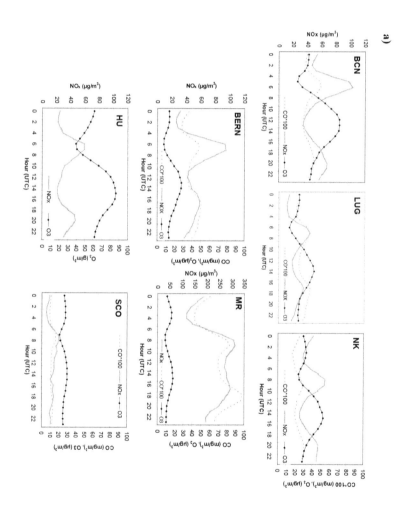

FIGURE 2A: Daily cycle of: gaseous pollutants concentrations (CO, NOx and O3) levels for each monitoring site

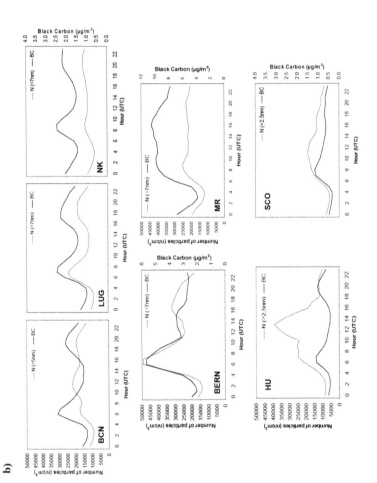

FIGURE 2B: Daily cycle of: Black Carbon (BC) and Number concentration (N) levels for each monitoring site.

FIGURE 2C: Daily cycle of: BC and PM10 levels for each monitoring site.

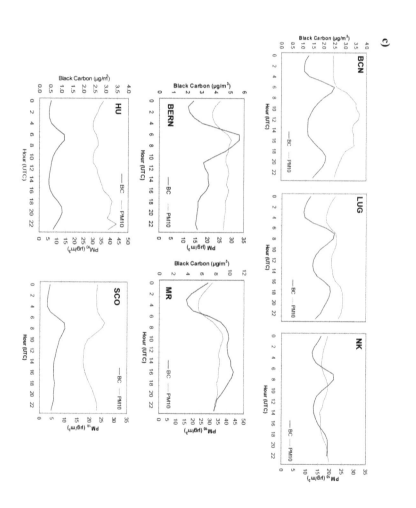

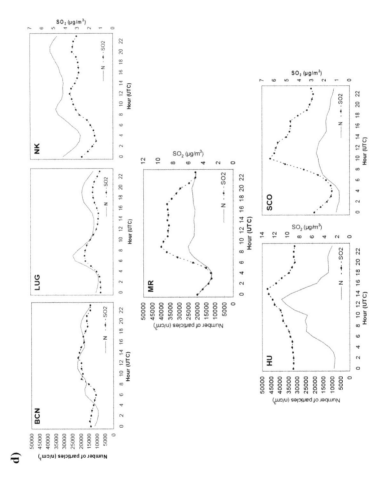

FIGURE 2D: Daily cycle of: SO2 levels for each monitoring site.

On the other hand, the variability of CO/BC ratio does not reflect the composition of the vehicle fleet in each city. Maximum values are recorded for BCN (235), while minimum values are obtained in MR (90). In LUG, NK, Bern and SCO values ranged between 143 and 188.

NO_2 levels reached 16–42 µg m^{-3} at the urban background environments, with higher concentrations in BCN, probably related to the higher percentage of diesel vehicles (with enhanced primary NO_2 and NOx emissions) in fleet, higher O_3 levels (leads to faster NO oxidation to NO_2) and the very elevated car density (6100 cars km^{-2}). At the traffic sites, NO_2 concentrations ranged from 24–127 µg m^{-3}, with the highest levels being recorded at MR, with the highest traffic flow (80 000 vehicles day^{-1}) and probably also to a significant proportion of diesel in the vehicle fleet of London (Carslaw et al., 2005). In fact, the ratio NOx/CO increases with the proportion of diesel vehicles and is very similar in BCN and NK. In SCO, NO_2 concentrations reach relatively low levels (7 µg m^{-3}) due to the good ventilation conditions on the island, NO_2 levels mostly represent the primary emissions.

Marked differences in the NO_2/NO ratio are only found between traffic (0.9–1.2) and urban background sites (2–3). These values are in the order of those obtained in other sites in Europe (Chaloulakou et al., 2007) and seem to be independent of the percentage of diesel vehicle in the fleet. Thus, despite certain evidences described above, differences in pattern emissions of gasoline and diesel vehicles are not clearly shown in the stations under study. It should be noted that NO_2 measurements can be overestimated because of interferences of oxidized nitrogen compounds in the conventional instruments equipped with molybdenum converters (Steinbacher et al., 2007).

Finally, O_3 concentrations range between 26 and 49 µg m^{-3} at urban background stations, including on the island. Very high values are measured at BCN, as a result of an intense photochemical activity. At traffic sites, values are about 13–16 µg m^{-3} as a consequence of a major consumption of O_3 by NO. Levels at HU were the highest (61 µg m^{-3}) as a direct consequence of high industrial emissions and solar radiation intensity.

10.4.2 TEMPORAL VARIABILITY OF ATMOSPHERIC POLLUTANTS

Figure 2a shows the 2009 average daily cycle of gaseous pollutants for each site. For all sites levels of NOx (NO_2 + NO) and CO follow the diurnal pattern of traffic intensity, reaching a maximum during the morning rush hour (07:00– 09:00 UTC), decreasing during the day because of atmospheric dilution processes, and increasing again in the evening (17:00-20:00 UTC). An increment of the ratio CO/NOx is observed comparing the evening peak with the one recorded at morning rush-hours. This phenomenon is more marked in BCN and Bern but the tendency is equal for all the stations. While CO emissions are a tracer of gasoline vehicles, NOx reflects emissions from diesel vehicles, so a possible explanation of the tendency above is the major presence of delivery HDV and LDV and of school buses in the morning.

The evolution of O_3 levels shows a typical diurnal pattern, with an increase at midday coinciding with the maximum photochemistry and vertical dilution. At SCO O_3 daily patterns differ from the above trend, with levels at night similar to those registered at midday, a behaviour induced by the continuous supply of fresh oceanic air masses coupled with low local NO levels.

The same daily evolution described for NOx and CO is followed by levels of BC (Fig. 2a and b), except in the case of Bern, where the second peak of BC is produced earlier than expected. The cause for the anomalous hourly trend detected in Bern is still unclear, although it is possibly related to the fact that the BC and the gaseous pollutant monitors were not co-located, but instead were distant by approximately 150 m on opposite sides of the road. The BC monitor was located closer to road traffic and to a railway. In all the other monitoring stations, BC levels traced accurately the impact of vehicle exhaust emissions on air quality, increasing with the proximity to roads, the density of vehicles, and the traffic flow. One may think that there is no need to measure BC levels if this aerosol component follows the variability of CO, NO and NO_2, however, differences among sites in the ratios CO/BC (ranging from 90–235) and NO_2/BC (ranging from 7–32) indicate that BC should be measured.

The combination of PM_{10} and BC in urban areas potentially constitutes a useful approach for air quality monitoring (Fig. 2c). While BC daily cycle is mostly determined by vehicle exhaust emissions, PM_{10} concentrations at these sites are also governed by non-exhaust particulate emissions resuspended by traffic, by midday atmospheric dilution and by other non-traffic emissions (see LUG and NK patterns in Fig. 2c). PM_{10} levels at the traffic sites remained nearly constant from the morning until the evening peak due to the effects of resuspension processes. In the case of BCN, concentrations increase at midday when sea breezes transport the re-suspended mineral material from the city towards the monitoring site. Similar results were reported in earlier studies (Querol et al., 1998; Harrison et al., 2001; Querol et al., 2001, 2005; Charron and Harrison, 2005). In contrast, PM10 concentrations in HU reach the highest values at night due to the seaward transport of aged particulate pollutants. During daylight, winds blow inland from the Atlantic Ocean carrying emission plumes with gaseous pollutants from industrial estates (Sanchez de la Campa et al., 2007), accounting for the different daily cycle of PM and gaseous pollutants.

N is also an appropriate tracer of traffic emissions in certain environments, but it has been reported to be highly influenced by photochemically induced nucleation (Pey et al., 2008; Perez et al., 2010; Fernandez-Camacho et al, 2010; Cheung et al., 2010). Peaks of N and BC at morning and afternoon rush-hours (07:00–09:00 and 17:00-20:00 UTC) are coincident in all the stations studied, with N being mainly influenced by primary aerosols and by the formation of new particles during the dilution and cooling of the vehicle exhaust emissions (Mariq et al., 2007, Wehner et al., 2009). Furthermore, at BCN, HU and SCO, N shows a second peak at midday, simultaneously occurring with the BC decrease, confirming that this peak could not be a consequence of primary emissions from road traffic, but of secondary formation of particles by means of photochemical nucleation processes from gaseous precursors. This midday nucleation takes place as a consequence of the high solar radiation, the growth of the mixing layer, the increase in wind speed and the consequent decrease of pollutants concentrations. This phenomenon is not observed in the selected northern and central European cities, where the decrease of N at midday was in the order of that of BC. The occurrence of nucleation events at midday in BCN was supported by means of an SMPS (scanning mobility particle sizers) working during the international DAURE campaign in 2009 (http://cires.colorado.edu/jimenez-group/wiki/index.php/DAURE) when it was observed that the increment of N at midday was caused by a marked increment of nucleation mode particles (N_{5-20}) (Fig. S4). Because of the similar pattern of N and meteorological parameters (global radiation, wind speed, wind direction and boundary layer), it was estimated that results regarding nucleation episodes from Barcelona could be extrapolated to these sites in southEurope.

In the case of traffic sites, daily patterns of BC and N follow an opposite trend since 11:00 UTC, so that the peaks of N in the evening show slight delays. The opposed trends of BC and N are specially significant in Bern, where the second peak of BC is registered earlier than in the rest of stations and is about 40 % lower than morning peak. This lack of parallelism between N and BC is due to a source of N other than traffic, which may be interpreted as secondary particles formed in the evening, when lower temperatures and lower mixing heights occur. The different location of the instrumentation can also be responsible for the dissimilarities. Sulphuric acid has been identified as a key nucleating substance in the atmosphere (Curtis et al., 2006) as atmospheric OH concentrations correlate well with UV solar radiation (Rohrer and Berresheim, 2006). Because UV radiation has not been measured continuously at most of the stations, we use the global intensity of solar radiation (SR) as its proxy with the purpose of obtaining the product of solar radiation (W m^{-2}), SO_2 (μg m^{-3}), O_3 (μg m^{-3}) and H_2Ov (g Kg^{-1}). This product was used as a surrogate parameter for H_2SO_4 production. Mean values of the product at midday (11:00-14:00 UTC) were 171 Wμg^2 m^{-8} in NK, 160 in MR, 107 in LUG, 1103 in BCN, 5271 in HU and 1259 in SCO, confirming the potential for secondary formation of particles at midday in the southern European cities under study.

Daily trends of SO_2 levels in BCN, HU and SCO suggest a significant source of this pollutant different from vehicle exhaust emissions (Fig. 2d). SO_2 levels in BCN attain a maximum from 9:00 to 13:00 UTC and around 10:00 UTC in SCO. At this time, breezes drive harbour emissions across the city, and thus in these cases this pollutant may be attributed to shipping emissions. In HU, the northward inland breeze blowing during the afternoon favours the inland transport of the SO_2 plumes over the city and the mixing of urban and industrial pollutants (Fernandez-Camacho et al., 2010). In LUG and MR, SO_2 concentrations are directly dependent upon exhaust emissions, reaching higher values with increases in traffic volume. In NK the diurnal cycle is also highly influenced by traffic, but maximum levels are registered at midday due probably to the convective mixing of upper tropospheric layers, polluted with SO_2 from power plants (Bigi and Harrison, 2010).

On a weekly scale, the study of the daily evolution of N and BC for each day of the week and each station (Fig. 3) shows that the daily evolution was not the same at weekends because the morning road traffic maximum disappears, and a relatively smoother daily evolution in aerosol concentrations during daylight was observed. Lower PM, N and BC levels at weekends are responsible for the sharper peak of N at midday registered for BCN, HU and SCO (especially marked on Sundays) due to the favourable conditions (low atmospheric pollution) for secondary aerosol formation by nucleation processes to take place. Low pollutant concentrations hinder condensation and coagulation processes but favour nucleation activated by photochemistry.

Furthermore, a seasonal trend for BC and N levels is not detected in stations with a direct influence of traffic (Fig. 4), with levels remaining relatively steady during the whole year. However, at LUG, NK and HU, concentrations are much lower in summer, coinciding with the higher dispersive conditions. By contrast, SCO accounts for the maximum values in summer probably due to major maritime traffic intensity. As expected, the N peaks at midday registered in BCN, HU and SCO, compared with the hourly average, are higher in summer.

CO and NOx concentrations follow the same seasonal trend described above for BC and N (Fig. S1). However, NOx declines in summer are also evident in BCN. As expected, O_3 levels present an opposite tendency, increasing during the months with the highest solar radiation intensities (Fig. S2). In BCN and HU, those periods with maximum concentrations of O_3 match with those of maximum concentrations of N at midday. This fact highlights the direct dependence of O_3 and N on photochemistry.

At SCO, O_3 levels are associated with processes of long-range transport and they show the typical spring maximum of the subtropical latitudes (Oltmans and Levy, 1994).

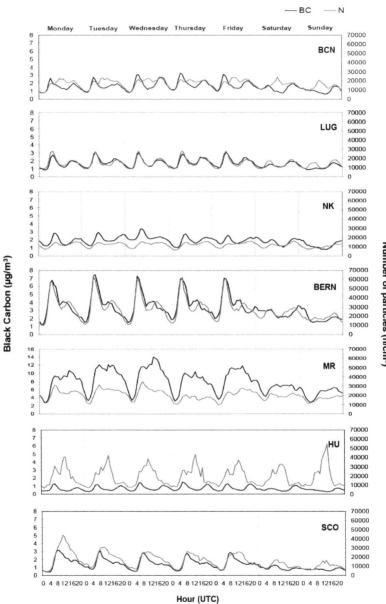

FIGURE 3: Daily cycle of: BC and N for each day of the week

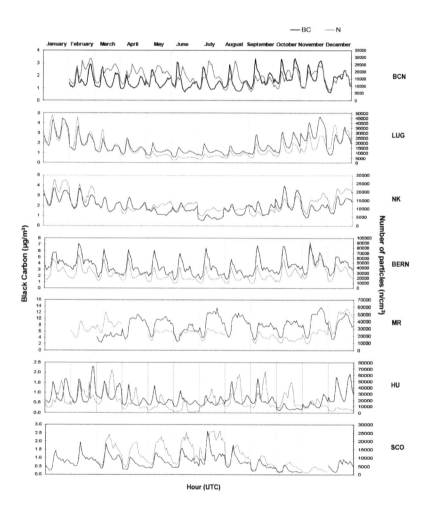

FIGURE 4: Daily cycle of: BC and N for each month.

10.4.3 BC/N RELATIONSHIP: SIMILARITIES AND DIFFERENCES ACROSS EUROPE

Parallelisms between BC and N are repeatedly observed in urban environments owing to the impact of vehicle exhaust emissions (Zhu et al., 2002; Fruin et al., 2004; Rodrıguez and Cuevas, 2007; Perez et al., 2010). The degree of correlation of these two parameters has been studied for each site using the methodology described by Rodriguez and Cuevas (2007). Different behaviours were observed depending on the time of the day.

BC vs. N scatter plots were analysed for four different periods of the day (Fig. S3), specifically: 07:00–09:00, 11:00–14:00, 18:00–21:00, 01:00–03:00 UTC. The selection of these time ranges is determined by variations in pollutant levels, mainly governed by emission patterns and atmospheric dynamics. The first period stands for the traf- fic rush-hours, accounting for the highest exhaust emissions. At 11:00–14:00 UTC there is an increase in the height of the mixing layer and the highest solar radiation intensity is reached, with the consequent development of mountain and sea breezes. These factors result in a dilution of atmospheric pollutants. The 18:00–21:00 UTC period represents the evening traffic rush hours, with pollutant concentrations in the order of those registered in the morning rush-hours, this period is also characterized by the influence of biomass burning emissions during winter, specially significant at north and central European locations. Finally, the night period is characterized by the lowest traffic intensity; however, the decrease of the height of the mixing layer causes a concentration of pollutants.

At any time and for all the stations, the N versus BC scatter plots are grouped between two defined lines with slopes S1 and S2 representing the minimum and maximum N/BC ratios, respectively (Fig. S3). S1 is interpreted as the minimum number of primary particles arising from vehicle exhaust emissions per each nanogram of ambient air BC. The observed increases in N/BC ratios up to reach the maximum S2 value are caused by means of enhancements in the new particle formation rates during the dilution and cooling of the vehicle exhaust emissions and/or in ambient air.

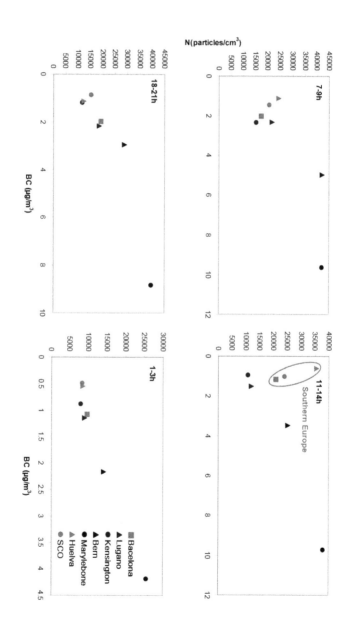

FIGURE 5: Correlation between: (a) the mean annual hourly levels of N and BC for the periods 07:00–09:00, 11:00–14:00, 18:00–21:00 and 01:00–03:00 (UTC) for the different study sites.

This analysis of the relationship between BC and N is performed assuming that BC in urban areas is an accurate tracer of primary traffic emissions. A similar approach was used by Turpin and Huntzicker (1995) to differentiate primary and secondary organic carbon

Important differences in S1 are observed between sites at 11:00–14:00 UTC (Table 5). It is important to note that dissimilarities are partly caused by the selection of CPC model at each location, since higher the cut size of the instrument is, lower the N/BC ratio will be. S1 ranges $3.1–4.1\times10^6$ particles ng^{-1} BC in central and north European cities, with the lowest values registered at the traffic sites. The ratio, expressed in 10^6 particles ng^{-1} BC, increases to 6.5 in BCN, 15.6 in SCO and 28.4 in HU. South European cities recorded the highest solar radiation just occurring when the atmospheric dynamics result in the highest dilution of pollutants (midday). This dilution is enhanced by the development of sea breezes in coastal cities. The combination of these factors sets a favourable scenario for nucleation events to occur. The formation of secondary particles by nucleation might account for the marked increase of S1 from 07:00–09:00 UTC to 11:00–14:00 UTC in BCN, HU and SCO. Furthermore, the influence of shipping emissions in these cities, together with an important industrial source in HU, adds the presence of gaseous precursors (mainly SO_2) to the conditions that favour nucleation.

As expected, S1 at 18:00–21:00 UTC is in the order of those ratios at 07:00–09:00 UTC (5.7×10^6 particles ng^{-1} BC in BCN, 8.7×10^6 particles ng^{-1} in HU, 11.1×10^6 particles ng^{-1} BC in SCO and $3.1–5.3\times10^6$ particles ng^{-1} BC in the rest of stations), although values are slightly higher at 18–21 h at most of the sites, probably due to a major gas to particle transfer of matter by condensation and/or nucleation. In MR, the N/BC ratios measured at 7–9 h and at 18:00–21:00 h are very similar, highlighting the lower relevance of secondary aerosol at this site, with persistently high exhaust emissions.

At night (01:00–03:00 UTC), the ratio ranges between $3.4–4.7$ 10^6 particles ng^{-1} BC for most of the sites, increasing to 8.3 10^6 particles ng^{-1} BC in HU and 9.2 10^6 particles ng^{-1} BC in SCO. In spite of the traffic intensity reduction, the decrease of the height of the mixing layer promotes condensation processes, resulting in similar N/BC ratios as those calculated at traffic rush-hours.

Figure 5 shows the correlations between the mean annual hourly levels of N and BC for the above S1 ratios (Table 5) during the periods 07:00–09:00, 11:00–14:00, 18:00–21:00 and 01:00–03:00 UTC for the different sites. It is evident that diurnal (07:00–09:00 and 11:00-14:00) N-BC patterns for southern Europe differ markedly from the general correlation found for central Europe, with higher N/BC trend. Thus, during these diurnal periods the cities from central Europe and UK are aligned along a very well fitted regression line between BC and N. The position of these cities (black and blue symbols) along the line is dependent on the traffic flow and proximity to roads as well as meteorological dispersion, with a clear positive slope. However, the southern European cities (red symbols) are well off this regression line. During some periods there seems to even be a negative slope. During the nocturnal periods both northern-central and southern European cities seem to fit the same regression line with a positive slope.

As regards S2, the observed trends are almost the same as those described for S1. However, the enhancement of N/BC at midday in BCN, HU and SCO is more marked when considering S2. In Bern, an increase of S2 is produced in the afternoon (18:00–21:00 UTC), since the peak of BC is produced earlier than expected. Such is the case of LUG, where the maximum ratio at 18:00–21:00 and mainly at 01:00–03:00 UTC is higher than in the rest of central and north European sites. This could be attributed to the influence of biomass burning emissions, since volatile organic compounds (VOCs) from residential wood combustion may lead to significant secondary non-fossil organic aerosols in winter as suggested to be important in Zurich (Lanz et al., 2008) and in Roveredo (Lanz et al., 2010).

Similar ratios have been reported in previous studies in Milano (Rodriguez and Cuevas, 2007), where the new particle formation contributions result in a strong correlation between the diurnal cycle of N/BC ratio and the solar radiation intensity.

The results indicate that BC is an appropriate tracer for the intensity and proximity of traffic emissions, and as a suitable indicator of the dynamics of the atmospheric mixing layer. In central and north European cities, N daily trends are similar to those of BC, whereas in south Europe, N is highly influenced by the occurrence of nucleation processes induced by photochemistry at midday.

10.4.4 DIFFERENCES IN PRIMARY EMISSIONS AND NUCLEATION ENHANCEMENTS ACROSS EUROPEAN CITIES

In order to quantify the sources and processes contributing to UFP, particle number concentration data were analysed using the methodology described by Rodrıguez and Cuevas (2007), where:

$$N1 = S1 \times BC \tag{2}$$

$$N2 = N - N1, \tag{3}$$

with N being the total number concentration; N1 the minimum primary emission of vehicle exhaust; N2 accounting for: (a) secondary particles formed in the atmosphere by homogeneous nucleation or other heterogeneous reactions from gaseous precursors arising from traffic or any urban source, (b) primary particles from other sources different to traffic such as biomass burning, resuspension, residential emissions and biogenic emissions, and (c) inherited particles present in the air mass, which receives anthropogenic emissions in a given point in time; BC representing the Black Carbon concentrations; and S1 the minimum N vs. BC slope observed during the morning rush hours (07:00–09:00 UTC). Values of S1 (expressed as 106 particles/ngBC) of 3.2 at LUG, 3.6 at NK and Bern, 2.9 at MR, 5.1 at BCN, 8.7 in HU and 9.9 in SCO were used. These differences in the S1 values are caused by: 1) the use of different CPC models (with different cut sizes) at different sites (the lowest S1 values are observed at the sites where the CPC with the largest cut size, 7 nm, was used) and (2) the influence of the ambient air conditions on the new particle formation during dilution and cooling of the vehicle exhaust.

Figure 6 shows hourly average values of N1 and N2 for every day of the week. Differences between sites are marked, observing two different patterns: stations with parallel cycles of N1 and N2 at midday (LUG and Bern) and stations with a clear decoupling of the two components, indicating no dependence on exhaust emissions of N2 at midday (BCN, HU,

SCO, NK and MR). It is a consequence of the described nucleation processes favoured by photochemistry.

Nonetheless, an increase of N2 not coinciding with N1 is also observed in NK possibly related with the midday SO_2 increase from power plants, around two hours before (12:00 UTC) as a result of the downward mixing upper tropospheric layers (Fig. 2d). This association between increases of N and power plant emissions has been observed in previous studies (Brock et al., 2002).

Table 6 lists the total average percentages of N1 and N2 and the average percentage at midday (11:00-14:00 UTC) on an hourly basis for each station. N1 (minimum primary emission of vehicle exhaust) accounts for 46, 38 and 46 % of the total N during the whole sampling period in BCN, HU and SCO, respectively. These percentages are quite similar to the one registered in LUG in winter (38 %), significantly lower than the percentage obtained in summer (53 %), maybe as a direct consequence of SOA formation from biomass burning emissions. The value increases to 54 % in NK, 45 % in Bern and to 78 % in MR, as a consequence of the important and consistent traffic impact. In south European cities, a 64 to 85 % N2 contribution is observed at midday (11:00–14:00 h UTC, representing secondary parts from gaseous precursors, primary parts from non-traffic sources, and/or particles inherited in the air mass). As previously reported this high N2 load is caused by the combination of 3 processes: (1) increased solar radiation, (2) the dilution of pollutants in an increased boundary layer height, and (3) the input of SO_2 from a source different than traffic exhaust emissions. These high N2 loads are also obtained in the same cities on a mean daily basis (66–80 %) during Sundays, mainly due to the lower levels of atmospheric pollutants.

Considering the relatively high concentrations of SO_2 at MR (with an average concentration at midday about 70 % higher than those registered in BCN and SCO) and the very high primary N load obtained, it can be stated that higher concentrations of SO_2 are not enough to achieve an enhancement of the secondary N load, but the combination with other factors such as solar radiation intensity and dilution (favouring decreases in PM levels) appears to be essential. This is supported by the observation of frequent nucleation events at the rural Harwell (UK) site with lower SO_2 concentrations (Charron et al., 2007).

To further confirm this hypothesis, Spearman rank correlation (ρ) tests were used to assess the relationship between N2 and different factors, specifically: N2 vs. SO_2* solar radiation intensity, N2 vs. wind velocity and N2 vs. wind direction on an hourly basis. Positive correlations between N2 and SO_2* solar global radiation were obtained in HU ($\rho = 0.86$), BCN ($\rho = 0.67$) and SCO ($\rho = 0.78$) with a level of significance of 0.01. The coefficient for the same test resulted lower than 0.20 in north and central European sites. No significant correlations were found in any case when analysing wind components. Nevertheless, it is known that enhancements in solar radiation intensity directly promote the development of sea breezes. This change in wind direction and velocity, coupled with increases in the mixing layer high, causes the dilution of atmospheric pollutants, favouring nucleation processes (Ketzel et al., 2004; Lee et al., 2008).

According to the results obtained it is clearly evidenced that N variability in different European urban environments is not equally influenced by the same emission sources and atmospheric processes.

10.5 CONCLUSIONS

This study shows the results of the interpretation of the 2009 variability of levels of PM, Black Carbon (BC), particle number concentration (N) and a number of gaseous pollutants at seven selected urban air quality monitoring sites covering road traffic, urban background, urban-industrial, and urbanshipping environments, from southern, central and northern Europe.

The results show that the variations of PM and N levels do not always reflect the variation of the impact of road traffic emissions on urban aerosols. However, BC levels vary proportionally to those of traffic related gaseous pollutants, such as CO, NO_2 and NO. Due to this high correlation, one may suppose that monitoring the levels of these gaseous pollutants would be enough to extrapolate exposure to BC levels. However the BC/CO, BC/NO_2 and BC/NO ratios vary widely among the studied cities, as a function of the distance to traffic emissions, the vehicle fleet composition and the influence of other emission sources such as biomass burning. Thus, BC is a relevant indicator for the impact of anthropogenic emissions at a measurement site and should therefore be measured in air quality monitoring networks. The combination of PM_{10} and BC in urban areas potentially constitutes a useful approach for air quality monitoring. Thus, BC is governed by vehicle exhaust emissions, while PM_{10} concentrations at these sites are also governed by non-exhaust particulate emissions resuspended by traffic, by midday atmospheric dilution and by other non-traffic emissions.

A subsequent question is focused on the evaluation of the variability of levels of N and the comparison with those of BC. The results indicate a narrow variation of primary road traffic N/BC ratios during traffic rush hours, while a wide variation of this ratio was determined for the noon period. Although in central and northern Europe N and BC levels tend to vary simultaneously, not only during the traffic rush hours but also during the whole day, in southern Europe maximum N levels coinciding with minimum BC levels are usually recorded at midday. These N maxima recorded in southern European urban background environments are attributed to midday nucleation episodes occurring when gaseous pollutants are diluted and maximum insolation and O_3 levels occur. The occurrence of SO_2 peaks may also contribute to the incidence of midday nucleation burst in specific industrial or shipping influenced areas, although at several central European sites similar levels of SO_2 are recorded without yielding nucleation episodes.

According to the results obtained it is clearly evidenced that N variability in different European urban environments is not equally influenced by the same emission sources and atmospheric processes. Therefore, we conclude that N variability does not always reflect the impact of road traffic on air quality in southern Europe, whereas BC is a more consistent tracer of such influence. However, N should be measured since ultrafine particles (<100 nm) may have large impacts on human health based on the very fine grain size that may reach the cardiovascular and cerebrovascular systems and the potential toxicity (Perez et al., 2009).

REFERENCES

1. Alonso-Perez, S., Cuevas, E., Querol, X., Viana, M., and Guerra, ´ J.C.: Impact of the Saharan dust outbreaks on the ambient levels of total suspended particles

(TSP) in the marine boundary layer (MBL) of the Subtropical Eastern North Atlantic Ocean, Atmos. Environ., 41, 9468–9480, 2007.

2. Anderson, H. R., Bremner, S. A., Atkinson, R. W., Harrison, R. M., and Walters, S.: Particulate matter and daily mortality and hospital admissions in the west midlands conurbation of the United Kingdom: associations with fine and coarse particles, black smoke and sulphate, Occup. Environ. Med., 58, 504–510, 2001.

3. Baldasano, J. M., Plana, J., Gonc‚alves, M., Jimenez, P., Jorba, O., and Lopez, E: Mejora de la calidad del aire por cambio de combustible a gas natural en automocion: Aplicaci on a Madrid y Barcelona, Fundacion Gas Natural, Spain, 2007. ´

4. Bigi, A. and Harrison, R. M.: Analysis of the air pollution climate at a central urban background site, Atmos. Environ., 44, 2004– 2010, 2010.

5. Boy, M. and Kulmala, M.: Nucleation events in the continental boundary layer: Influence of physical and meteorological parameters, Atmos. Chem. Phys., 2, 1–16, doi:10.5194/acp-2-1-2002, 2002.

6. Brock, C. A., Washenfelder, R. A., Trainer, M., Ryerson, T. B., Wilson, J. C., Reeves, J. M., Huey, L. G., Holloway, J. S., Parrish, D. D., Hubler, G., and Fehsenfeld, F. C.: Particle growth in the plumes of coal-fired power plants, J. Geophys. Res., 107, AAC9, D12, 4155, doi:10.1029/2001JD001062, 2002. 2002.

7. Bukowiecki, N., Dommen, J., Prev´ ot, A. S. H., Weingartner, E., and Baltensperger, U.: Fine and ultrafine particles in the Zrich (Switzerland) area measured with a mobile laboratory: an assessment of the seasonal and regional variation throughout a year, Atmos. Chem. Phys., 3, 1477–1494, doi:10.5194/acp-3-1477-2003, 2003.

8. Carslaw, D.: Evidence of an increasing NO2/NOx emissions ratio from road traffic emissions, Atmos. Environ., 39, 4793–4802, 2005.

9. Casati, R., Scheer, V., Vogt, R., and Benter, T.: Measurements of nucleation and soot mode particle emission from a diesel passenger car in real world and laboratory in situ dilution, Atmos. Environ., 41, 2125–2135, 2007

10. Cavalli, F., Viana, M., Yttri, K. E., Genberg, J., and Putaud, J.- P.: Toward a standardised thermal-optical protocol for measuring atmospheric organic and elemental carbon: the EUSAAR protocol, Atmos. Meas. Tech., 3, 79–89, doi:10.5194/amt-3-79-2010, 2010.

11. Charron, A. and Harrison, R. M.: Fine (PM2.5) and coarse (PM2.5−10) particulate matter on a heavily trafficked London highway: sources and processes, Environ. Sci. Technol., 39, 7768–7776, 2005.

12. Charron, A., Birmili, W., and Harrison, R. M.: Factors influencing new particle formation at the rural site, Harwell, United Kindom, J. Geophys. Res., 112, D14210, 15, doi:10.1029/2007/JD008425, 2007,

13. Cheung, H. C., Morawska, L., and Ristovski, Z. D.: Observation of new particle formation in subtropical urban environment, Atmos. Chem. Phys., 11, 3823–3833, doi:10.5194/acp-11-3823-2011, 2011.

14. Chio, C.-P. and Liao, C.-M.: Assessment of atmospheric ultrafine carbon particle-induced human health risk based on surface area dosimetry, Atmos. Environ., 42, 8575–8584, 2008.

15. Curtis, J.: Nucleation of atmospheric aerosol particles, C. R. Physique, 7, 1027–1045, 2006.

16. Dall'Osto, M., Thorpe, A., Beddows, D. C. S., Harrison, R. M., Barlow, J. F., Dunbar, T., Williams, P. I., and Coe, H.: Remarkable dynamics of nanoparticles in the urban atmosphere, Atmos. Chem. Phys. Discuss., 10, 30651–30689, doi:10.5194/acpd-10- 30651-2010, 2010.

17. Department of Transport, UK: http://www.dft.gov.uk, last access: February 2011.

18. Direccion General de Trafico, Spain: ' http://apl.dgt.es/IEST2, last access: February 2011.

19. Dunn, M. J., Jimenez, J. L., Baumgardner, D., Castro, T., Mc- ' Murry, P. H., and Smith, J. N.: Measurements of Mexico City nanoparticle size distributions: Observations of new particle formation and growth, Geophys. Res. Lett., 31, L10102, doi:10.1029/2004GL019483, 2004.

20. Ecoplan (2007): Auswertung Mikrozensus 2005 fur den Kan- " ton Bern, Report, Bern: http://www.bve.be.ch/bve/de/index/direktion/ueber-die-direktion/statistik. assetref/content/dam/documents/BVE/GS/de/Abteilung-Gesamtmobilit%C3%A4t Mikrozensus-Verkehrsverhalten-2005 Kt-BE.pdf, last access: February 2011.

21. Fernandez-Camacho, R., Rodr ' 'ıguez, S., de la Rosa, J., Sanchez de ' la Campa, A. M., Viana, M., Alastuey, A., and Querol, X.: Ultra-fine particle formation in the inland sea breeze airflow in Southwest Europe, Atmos. Chem. Phys. Discuss., 10, 17753–17788, doi:10.5194/acpd-10-17753-2010, 2010.

22. Fischer, P. H., Hoek, G., van Reeuwijk, H., Briggs, D. J., Lebret, E., van Wijnen, J. H., Kingham, S., and Elliott, P. E.: Traffic-related differences in outdoor and indoor concentrations of particles and volatile organic compounds in Amsterdam, Atmos, Environ, 34, 3713–3722, 2000.

23. Fruin, S., Westerdahl, T., Sax, C., Sioutas C., and Fine, P. M.: Measurements and predictors of on-road ultrafine particle concentrations and associated pollutants in Los Angeles, Atmos. Environ., 42, 207–219, 2008.

24. Gao, J., Wang, T., Zhou, X., Wu, W. and Wang, W.: Measurement of aerosol nu ber size distributions in the Yangtze River delta in China: Formation and growth of particles under polluted conditions, Atmos. Environ., 43, 829–836, 2009.

25. Guerra, J. C., Rodrıguez, S., Arencibia, M. T., and Garcıa, M. D.: Study on the formation and transport of ozone in relation to the air quality management and vegetation protection in Tenerife (Canary Islands), Chemosphere, 56, 1157–1167, 2004.

26. Hameri, K., Kulmala, M., Aalto, P., Leszczynski, K., Visuri, R., and " Hamekoski, K.: The investigation of aerosol particle formation " in urban background area of Helsinki, Atmos. Res., 41, 281–298, 1996.

27. Hamilton, R. S. and Mansfield, T. A.: Airborne particulate elemental carbon: its sources, transport and contribution to dark smoke and soiling, Atmos. Environ., 25, 715–723, 1991.

28. Harrison, R. M., Deacon, A. R., Jones, M. R., and Appleby, R. S.: Sources and processes affecting concentrations of PM10 and PM2.5 particulate matter in Birmingham (UK), Atmos. Environ., 31, 4103–4117, 1997.

29. Harrison, R. M., Yin, J., Mark, D., Stedman, J., Appleby, R. S., Booker, J., and Moorcroft S.: Studies of the coarse particle (2.5– 10μm) component in UK urban atmospheres, Atmos. Environ., 35, 3667–3679, 2001.

30. Harrison, R. M., Jones, A. M., and Lawrence, R. G.: Major component composition of PM10 and PM2.5 from roadside and urban background sites, Atmos. Environ., 38, 4531–4538, 2004.

31. Hueglin, C., Buchmann, B., and Weber, R. O.: Long-term observation of real-worls road traffic emision factors on a motorway in ' Switzerland, Atmos. Environ, 40, 3696–3709, 2006.

32. Imhof, D., Weingartner, E., Pr ¨ ev´ ot, A. S. H., Ordo ˆ nez, C., Kurten- ˜ bach, R., Wiesen, P., Rodler, J., Sturm, P., McCrae, I., Ekstrom, M., and Baltensperger, U.: Aerosol and NOx emission factors and submicron particle number size distributions in two road tunnels with different traffic regimes, Atmos. Chem. Phys., 6, 2215–2230, doi:10.5194/acp-6-2215-2006, 2006.

33. Janssen, N. A. H., Van Mansom, D. F. M., Van Der Jagt, K., Harssema, H., and Hoek, G.: Mass concentration and elemental composition of airborne particulate matter at street and background locations, Atmos. Environ., 31, 1185–1193, 1997.

34. Johnson, G. R., Ristovski, Z. D., Anna, B. D., and Morawska, L.: The hygroscopic behaviour of partially volatilized coastal marine aerosols using the VH-TDMA technique, J. Geophys. Res., 110, D20203, doi:10.1029/2004JD005657, 2005.

35. Ketzel, M., Wahlin, P., Kristensson, A., Swietlicki, E., Berkowicz, ° R., Nielsen, O. J., and Palmgren, F.: Particle size distribution and particle mass measurements at urban,near-city and rural level in the Copenhagen area and Southern Sweden, Atmos. Chem. Phys., 4, 281–292, doi:10.5194/acp-4-281-2004, 2004.

36. Kittelson, D. B.: Engines and nanoparticles: a review, J. Aerosol Sci., 29, 575–588, 1998.

37. Klemm, R. J. and Mason, R. M.: Aerosol Research and Inhalation Epidemiological Study (ARIES): air quality and daily mortality statistical modelling-interim results, J. Air Waste Manage., 50, 1433–1439, 2000.

38. Kulmala, M., Vehkamaki, H., Petaja, T., Dal Maso, M., Lauri, A., Kerminen, V.M., Birmili, W., and McMurry, P. H.: Formation and growth rates of ultrafine atmospheric particles: a review of observations, J. Aerosol Sci., 35, 143–175, 2004.

39. Lanz, V., Alfarra, A., Baltensperger, U., Burchmann, B., Huegling, C., Szidat, S., Wehrli, M. N., Wacker, L., Weimer, S., Caseiro, A., Puxbaum, H., and Prev´ ot, A. S. H.: Source attribution of submicron organic aerosols during wintertime inversions by advanced factor analysis of aerosol mass spectra, Environ. Sci. Technol., 42, 214–220, 2008.

40. Lanz, V. A., Prevot, A. S. H., Alfarra, M. R., Weimer, S., Mohr, C., DeCarlo, P. F., Gianini, M. F. D., Hueglin, C., Schneider, J., Favez, O., D'Anna, B., George, C., and Baltensperger, U.: Characterization of aerosol chemical composition with aerosol mass spectrometry in Central Europe: an overview, Atmos. Chem. Phys., 10, 10453–10471, doi:10.5194/acp-10-10453-2010, 2010.

41. Lee, Y. G., Lee, H. W., Kim, M. S., Choi, C. Y., and Kim, J.: Characteristics of particle formation events in the coastal region of Korea in 2005, Atmos. Environ., 42, 3729–3739, 2008.

42. Li, N., Sioutas, C., Cho, A., Schmitz, D., Misra, C., Sempf, J., Wan, M., Oberley, T., Froines, T., Nel, A. Ultrafine Particulate Pollutants Induce Oxidative Stress and Mitochondrial Damage, Environ. Health Persp., 111, 4–9, 2003.

43. Maricq, M. M.: Chemical characterization of particulate emissions from diesel engines: A review, Aerosol Sci., 38, 1079–1118, 2007.
44. Mejia, J. F. and Morawska, L.: An investigation of nucleation events in a coastal urban environment in the Southern Hemisphere, Atmos. Chem. Phys., 9, 7877–7888, doi:10.5194/acp-9-7877-2009, 2009.
45. Metzger, A., Verheggen, B., Dommen, J., Duplissy, J., Prevot, A. S. H., Weingartner, E., Riipinen, I., Kulmala, M., Spracklen, D. V., Carslaw, K. S., and Baltensperger, U.: Evidence for the role of organics in aerosol particle formation under atmospheric conditions, Proceedings of the National Academy of Sciences of the United States of America, 107(15), 6646–6651, 2010.
46. Minoura, H. and Takekawa, H.: Observation of number concentrations of atmospheric aerosols and analysis of nanoparticle behavior at an urban background area in Japan, Atmos. Environ., 39, 5806–5816, 2005.
47. Moore, K. F., Ning, Z., Ntziachristos, L., Schauer, J. J., and Sioutas, C.: Daily variation in the properties of urban ultrafine aerosolPart I: Physical characterization and volatility, Atmos. Environ., 41, 8633–8646, 2007.
48. Morawska, L., Thomas, S., Bofinger, N., Wainwright, D., and Neale, D.: Comprehensive characterization of aerosol in a subtropical urban atmosphere particle size distribution and correlation with gaseous pollutants, Atmos. Environ., 32, 2467–2478, 1998.
49. Morawska, L., Jayaratne, E. R., Mengersen, K., Jamriska, M., and Thomas, S.: Differences in airborne particle and gaseous concentrations in urban aor between weekdays and weekends, Atmos. Environ., 36, 4375–4383, 2002.
50. Morawska, L., Ristovski, Z., Jayaratne, E. R., Keogh, D. U., and Ling, X.: Ambient nano and ultrafine particles from motor vehicle emissions: Characteristics, ambient processing and implications on human exposure, Atmos. Environ, 42, 8113–8138, 2008.
51. Muller, T., Henzing, J. S., de Leeuw, G., Wiedensohler, A., ¨Alastuey, A., Angelov, H., Bizjak, M., Collaud Coen, M., Engstrom, J. E., Gruening, C., Hillamo, R., Hoffer, A., Imre, K., Ivanow, P., Jennings, G., Sun, J. Y., Kalivitis, N., Karlsson, H., Komppula, M., Laj, P., Li, S.-M., Lunder, C., Marinoni, A., Martins dos Santos, S., Moerman, M., Nowak, A., Ogren, J. A., Petzold, A., Pichon, J. M., Rodriquez, S., Sharma, S., Sheridan, P. J., Teinila, K., Tuch, T., Viana, M., Virkkula, A., Weingart- ner, E., Wilhelm, R., and Wang, Y. Q.: Characterization and intercomparison of aerosol absorption photometers: result of two intercomparison workshops, Atmos. Meas. Tech., 4, 245–268, doi:10.5194/amt-4-245-2011, 2011.
52. Nel, A.: Air pollution-related illness: effects if particles, Science, 308, 804–806, 2005.
53. Oltmans, S. J. and Levy II, H.: Surface ozone measurements from a global network, Atmos. Environ., 28, 9–24, 1994.
54. Pakkanen, T. A., Kerminen, V. M., Ojanena, C. H., Hillamo, R. E., Aarnio, P., and Koskentalo, T.: Atmospheric Black Carbon in Helsinki, Atmos. Environ., 34, 1497–1506, 2000.
55. Park, K., Park, J. Y., Kwak, J., Cho, G. N., and Kim, J.: Seasonal and diurnal variations of ultrafine particle concentration in urban Gwangju, Korea: Observation of ultrafine particle events, Atmos. Environ., 42, 788–799, 2008.

56. Perez, C., Sicard, M., Jorba, O., Comer ´ on, A., Baldasano, J. ´ M.: Summertime re-irculations of air pollutants over the northeastern Iberian coast observed from systematic EARLINET lidar measurements in Barcelona, Atmos. Environ., 38, 3983–4000, 2004.

57. Perez, L., Medina-Ram ´ on, M., K ´ unzli, N, Alastuey, A., Pey, J., Perez, N., Garcia, A., Tobias, A., Querol, X., and Sunyer, J.: ´ Size fractionated particulate matter, vehicle traffic, and casespecific daily mortality in Barcelona (Spain), Environ. Sci. Technol., 43(13), 4707–4714, doi:10.1021/es8031488, 2009.

58. Perez, N., Pey, J., Cusack, M., Reche, C., Querol, X., Alastuey, A., and Viana, M.: Variability of Particle Number, Black Carbon, and PM10, PM2.5, and PM1 Levels and Speciation: Influence of Road Traffic Emissions on Urban Air Quality, Aerosol Sci. Technol., 44, 487–499, 2010.

59. Petzold, A. and Schonlinnes, M.: Multi-angle absorption photometry-a new method for the measurement of aerosol light absorption and atmospheric Black Carbon, J. Aerosol Sci., 35, 421–441, 2004.

60. Pey, J., Rodr´ıguez, S., Querol, X, Alastuey, A., Moreno, T., Putaud, J. P., and Van Dingenen, R.: Events and cycles of urban aerosol in the western Mediterranean, Atmos. Environ., 42, 9052–9062, 2008.

61. Pey, J., Querol, X., Alastuey, A., Rodr´ıguez, S., Putaud, J. P., and Van Dingenen, R.: Source apportionment of urban fine and ultra- fine particle number concentration in a Western Mediterranean city, Atmos. Environ., 43, 4407–4415, 2009.

62. Querol, X., Alastuey, A., Puicercus, J.A., Mantilla, E., Miroa, J.V., Lopez-Soler, A., Plana, F., and Artinano, B. Seasonal evolution ¯ of suspended particles around a large coal-fired power station: particulate levels and sources, Atmos. Environ., 32, 1963–1978, 1998.

63. Querol, X., Alastuey, A., Rodriguez, S., Plana, F., Ruiz, R.C., Cots, N., Massague, G., and Puig, O.: PM10 and PM2.5 source apportionment in the Barcelona Metropolitan area, Catalonia, Spain, Atmos. Environ., 35, 6407–6419, 2005.

64. Qian, S., Sakurai, H., and McMurry, P. H.: Characteristics of regional nucleation events in urban East St. Louis, Atmos. Environ., 41, 4119–4127, 2008.

65. Rodr´ıguez, S. and Cuevas, E.: The contributions of "minimum primary emissions" and "new particle formation enhancements" to the particle number concentration in urban air, J. Aerosol Sci., 38, 1207–1219, doi:10.1016/j.jaerosci.2007.09.001, 2007.

66. Rodr´ıguez, S., Querol, X., Alastuey, A., Kallos, G., and Kakaliagou, O.: Saharan dust contributions to PM10 and TSP levels in Southern and Eastern Spain, Atmos. Environ., 35, 2433–2447, 2001.

67. Rodr´ıguez, S., Van Dingenen, R., Putaud, J. P., Martins-Dos Santos, S., and Roselli, D.: Nucleation and growth of new particles in the rural atmosphere of Northern Italy-relationship to air quality monitoring, Atmos. Environ., 39, 6734–6746, 2005.

68. Rodr´ıguez, S., Cuevas, E., Gonzalez, Y., Ramos, R., Romero, P. M., Perez, N., Querol, X., and Alastuey, A.: Influence of sea breeze ´ circulation and road traffic emissions on the relationship between particle number, Black Carbon, PM1, PM2.5 and PM2.5−10 concentrations in a coastal city, Atmos. Environ., 42, 6523–6534, DOI:10.1016/j.atmosenv.2008.04.022, 2008.

69. Rohrer, F. and Berresheim, H.: Strong correlation between levels of tropospheric hydroxyl radicals and solar ultraviolet radiation, Nature, 442(7099), 184–187, 2006.

70. Ronkko, T., Virtanen, A., Vaaraslahti, K., Keskinen, J., Pirjola, L. ¨ and Lappi, M.: Effect of dilution conditions and driving parameters on nucleation mode particles in diesel exhaust: Laboratory and on-road study, Atmos. Environ., 40, 2893–2901, 2006.

71. Sanchez de la Campa, A., de la Rosa, J., Querol, X., Alastuey, A., Mantilla, E.: Geochemistry and origin of PM10 in the Huelva region, Southwestern Spain, Environ. Res., 103, 305–316, 2007.

72. Sandradewi, J., Prevot, A. S. H., Weingartner, E., Schmidhauser, R., Gysel, M., Baltensperger, U.: A study of wood burning and traffic aerosols in an Alpine valley using a multi-wavelength aethalometer, Atmos. Environ., 42, 101–112, 2008.

73. Schwartz, S. E., Harshvardhan, and Benkovitz, C. M.: Influence of anthropogenic aerosol on cloud optical depth shown in satellite measurements and chemical transport modelling, Proc. Natl Acad. Sci., 99, 1784–1789, 2002.

74. Shi, Z., Shao, L., Jones, T. P., Whittaker, A. G., Richards, R. J., and Zhang, P.: Oxidative stress on plasmid DNA induced by inhalable particles in the urban atmosphere, Chinese Sci. Bull., 49, 692–697, doi:10.1007/BF03184267, 2004.

75. Smargiassi, A., Baldwin, M., Pilger, C., Dugandzic, R., and Brauer, M.: Small-scale spatial variability of particle concentration and traffic levels in Montreal: a pilot study, Sci. Total Environ., 338, 243–251, 2005.

76. Steinbacher, M., Zellweger, C., Schwarzenbach, B., Bugmann, S., Buchmann, B., Ordonez, C., Prevot, A.S.H., and Hueglin, C.: Nitrogen oxide measurements at rural sites in Switzerland: Bias of conventional measurement techniques, J. Geophys. Res., 112, D11307, doi:10.1029/2006JD007971, 2007.

77. Stolzel, M., Breitner, S., Cyrys, J., Pitz, M., Wolke, G., Kreyling, W., Heinrich, J., Wichmann, H. E., and Peters, A.: Daily mortality and particulate matter in different size classes in Erfurt, Germany, J. Exp. Sc. Environ. Epidemiol., 17, 458–467, 2007.

78. Swiss Federal Statistical Office: http://www.bfs.admin.ch, last access: February 2011.

79. Szidat, S., Prevot, A. S. H., Sandradewi, J., Alfarra, M. R., Synal, H.-A., Wacker, L. and Baltensperger U.: Dominant impact of residential wood burning on particulate matter in Alpine valleys during winter, Geophys. Res. Lett., 34, L05820, doi:10.1029/2006GL028325, 2007.

80. Tiresia: www.tiresia.ch, last access: February 2011.

81. Turpin, B. J. and Huntzicker, J. J.: Identification of secondary organic aerosol episodes and quantification of primary and secondary organic aerosol concentrations during SCAQS, Atmos.Environ., 29, 3527–3544, 1995.

82. Van Dingenen, R., Raes, F., Putaud, J.P., Baltensperger, U., Charron, A., Facchini, M.C., Decesari, S., Fuzzi, S., Gehrig, R., Hansson, H. C. Harrison, R. M., Huglin, C., Jones, A. M., Laj, ¨ P., Lorbeer, G., Maenhout, W., Palmgren, F., Querol, X., Rodriguez, S., Schneider, J., Brink, H., Tunved, P., Tørseth, K., Wehner, B., Weingartner, E., Wiedensohler, A., Whlin, P. : A European aerosol phenomenology – 1: physical characteristics of particulate matter at kerbside, urban, rural and background sites in Europe, Atmos. Environ., 38, 2561–2577, 2004.

83. Viana, M., Querol, X., Alastuey, A., Cuevas, E., andRodr´ıguez, S.: Influence of African dust on the levels of atmospheric particulates in the Canary Islands air quality network, Atmos. Environ., 36, 5861–5875, 2002.

84. Von Klot, S., Wolke, G., Tuch, T., Heinrich, J., Dockery, D.W., Schwartz, J., Krey-ling, W.G., Wichmann, H. E., andPeters, A.: Increased asthma medication use in association with ambient fine and ultrafine particles, Eur. Respir. J., 20, 691–702, 2002.
85. Watson, J. G., Chow, J. C., Lowenthal, D. H., Pritchett, L. C., Frazier, C. A., Neuroth, G. R., and Robbins, R.: Differences in the carbon composition of source profiles for diesel- and gasolinepowered vehicles, Atmos. Environ., 28(15), 2493–2505, 1994.
86. Weber, R. J., Marti, J. J., McMurry, P. H., Eisele, F. L., Tanner, D. J., and Jefferson, A.: Measurements of new particle formation and ultrafine particle growth rates at a clean continental site, J. Geophys. Res., D102, 4375–4385, 1997.
87. Wehner, B., Birmili, W., Gnauk, T., and Wiedensohler, A.: Particle number size dis-tribution in a street canyon and their transformation into the urban-air background: measurements and a simple model study, Atmos. Environ., 36, 2215–2223, 2002.
88. Wehner, B., Uhrner, U., von Lowis, S., Zallinger, M., and Wiedensohler, A.: Aerosol number size distributions within the exhaust plume of a diesel and a gasoline pas-senger car under on-road conditions and determination of emission factors, Atmos. Eviron., 43, 1235–1245, 2009.
89. Wikipedia: http://en.wikipedia.org/wiki/London, last access: January 2011, 2010.
90. Wichmann, H. E., Spix, C., Tuch, T., Wolke, G., Peters, A., Heinrich, J., Kreyling, W. G., and Heyder, J.: Daily Mortality and Fine and Ultrafine Particles in Erfurt, Germany, Part I: Role of Particle Number and Particle Mass, HE Publications, 98, 2000.
91. Zhang, K. M., Wexler, A. S., Zhu, Y. F., Hinds, W. C., and Sioutas, C.: Evolution of particle number distribution near roadways, Part II: the "Road-to-Ambient" process, Atmos. Environ., 38, 6655–6665, 2004.
92. Zhu, Y., Hinds, W. C., Kim, S., Shen, S., and Sioutas, C.: Study of ultrafine par-ticles near a major highway with heavy-duty diesel traffic, Atmos. Environ., 36, 4323–4335, 2002
93. Zong-bo, S., Ke-bin, H., Xue-chun, Y., Zhi-liang, Y., Fu-mo, Y., Yong-liang, M., Rui, M., Ying-tao, J., and Jie, Z.: Diurnal variation of number concentration and size distribution of ultrafine particles in the urban atmosphere of Beijing in winter, J. Environ. Sci., 19, 933–938, 2007.

Tables 3–6, as well as several supplemental files, are not available in this version of the article. To view this additional information, please use the citation on the first page of this chapter.

CHAPTER 11

Air Quality Modeling in Support of the Near-Road Exposures and Effects of Urban Air Pollutants Study (NEXUS)

VLAD ISAKOV, SARAVANAN ARUNACHALAM, STUART BATTERMAN, SARAH BEREZNICKI, JANET BURKE, KATHIE DIONISIO, VAL GARCIA, DAVID HEIST, STEVE PERRY, MICHELLE SNYDER, AND ALAN VETTE

11.1 INTRODUCTION

Studies of health effects associated with exposure to traffic-related air pollutants have typically used surrogates of exposure, such as residential proximity to roadways, traffic volumes on nearby roadways, and land-use regression techniques, to estimate exposure for the study population [1,2,3,4,5,6]. While these exposure metrics are relatively simple to generate and have minimal data requirements, they do not capture potentially important influences on spatial variability, and perhaps more importantly, temporal variability of traffic-related air pollutants such as factors that

Air Quality Modeling in Support of the Near-Road Exposures and Effects of Urban Air Pollutants Study (NEXUS). © *Isakov V, Arunachalam S, Batterman S, Bereznicki S, Burke J, Dionisio K, Garcia V, Heist D, Perry S, Snyder M, and Vette A.* International Journal of Environmental Research and Public Health *11,9 (2014). doi:10.3390/ijerph110908777. Licensed under a Creative Commons Attribution 3.0 Unported License.*

affect dispersion [7]. Traffic-related air pollutants can have significant temporal variability due to traffic activity patterns (e.g., rush hour peaks, higher during weekdays vs. weekends), emission profiles that vary with temperature, and the influence of meteorology, which are not captured by static exposure estimates based on geographic parameters (i.e., proximity to roadway, traffic intensity, lane use, etc.) that are often used in traffic studies.

Health studies of the effects of traffic-related pollutants have historically relied on exposure metrics such as those listed above because available central site measurements often do not adequately capture local influences from traffic. Data from regulatory monitoring sites may capture temporal variations for some pollutants (e.g., NOx, CO), but spatial coverage within an urban area is generally limited to one or two sites. Studies deploying multiple monitors to provide spatial coverage are costly, so samplers with lower temporal resolution (daily to weekly) are often used [8,9]. The spatial impact of traffic emissions also varies by pollutant due to their chemical and physical characteristics [10], therefore a number of different monitors are needed to obtain data for the various traffic-related air pollutants.

To address the limitations of available monitoring data and the various metrics of exposure, recent studies have utilized emission/dispersion models and daily activity locations to derive air pollution exposures for epidemiological studies [11,12,13,14,15,16,17]. Two main types of air quality models are relevant for this purpose: grid-based chemical transport models and plume dispersion models. Grid-based chemical transport models, such as the Community Multiscale Air Quality (CMAQ) model, estimate concentrations for large geographic areas at high time resolution but cannot resolve features smaller than a grid cell, usually several kilometers across [18]. Plume dispersion models, such as American Meteorological Society/Environmental Protection Agency Regulatory Model (AERMOD) [19], can provide locally resolved concentration gradients such as those occurring close to roadways but require estimates of background concentrations to compare model results to measurement data [20]. To account for the limitations of each type of model, a hybrid approach can be used where output from both a grid-based chemical transport model and a plume dispersion model are merged to provide contributions from photochemical interactions, long-range (regional) transport, and details attributable to local-scale dispersion [21,22].

The Near-road Exposures and Effects of Urban Air Pollutants Study (NEXUS) is investigating the respiratory health impacts of exposure to traffic-related air pollutants for children with asthma living near major roads in Detroit, MI [23]. Air quality modeling was included in the design of NEXUS to estimate exposure to traffic-related air pollutants that varied both spatially and temporally. Exposure estimates will be used for evaluating associations with daily health measurements collected during a 14-day period in each of four seasons for each study participant over a 1.5 years period. This paper describes application of the hybrid air quality modeling approach. The hybrid modeling components are described along with the specific inputs used for application to the Detroit study area and NEXUS participant locations. Model results are compared with available measurement data from regulatory monitoring sites within Detroit and intensive field studies conducted during NEXUS. The various exposure metrics produced from the model output which include the mobile source contribution to total exposure are provided for use in related NEXUS epidemiologic analysis, and described and compared here.

11.2 AIR QUALITY MODELING APPROACH FOR ESTIMATING EXPOSURE METRICS

We use a combination of local-scale dispersion models, regional-scale models and observations to provide temporally and spatially-resolved pollutant concentrations for the epidemiologic analysis. Local variations in emissions and meteorology were estimated using a combination of AERMOD and RLINE [24,25] dispersion models. RLINE is a research-level, line-source dispersion model developed by U.S. EPA's Office of Research and Development as a part of the ongoing effort to further develop tools for a comprehensive evaluation of air quality impacts in the near-road environment. This model incorporates traffic activity and primary mobile source emissions estimates to model hourly exposures to traffic emissions for the NEXUS participants. Exposures to air pollution from stationary sources such as manufacturing facilities and other non-road mobile sources were modeled using AERMOD. The input data including the source locations, emission rates, source parameters and other information were obtained from the 2008 official version of the National Emissions Inventory (NEI) from the U.S. EPA, the latest available at the time of the study [26].

To generate the total exposure of the NEXUS study participants, the urban background contribution must be added to the local estimates of exposure provided by AERMOD and RLINE models. The background contribution was estimated using a combination of the Community Multiscale Air Quality (CMAQ) model and the Space/Time Ordinary Kriging (STOK) model [27]. Two CMAQ model simulations were conducted: the baseline simulation represented all emissions in a broad region (covering the eastern US); the second removed all anthropogenic emissions in the NEXUS study domain. The ratios of concentrations predicted by CMAQ in these two simulations in the Detroit region along with measurements from the routine observational network in the region were used to estimate background pollutant concentrations at the NEXUS study locations.

The modeling provided hourly pollutant concentrations for CO, NOx, total $PM_{2.5}$ mass, and its components such as elemental carbon (EC) and organic carbon (OC), and benzene. Hourly concentrations were processed to calculate daily and annual average exposure metrics for each study participants' home and school location. The model-based exposure metrics provided the necessary inputs for use in the epidemiologic analyses to determine if children in Detroit, MI with asthma living in close proximity to major roadways have greater health impacts associated with traffic-related air pollutants than those living farther away, particularly for children living near roadways with high diesel traffic. Children were recruited on the basis of the proximity of their residence to roadways in three exposure groups: children living within 150 m of high traffic and high diesel (HD) roads, defined as having traffic that exceeds 6000 commercial vehicles/day (commercial annual average daily traffic; CAADT) and 90,000 total vehicles/day (annual average daily traffic; AADT); children living within 150 m of high traffic low diesel (LD) road, defined similarly but only including roads with CAADT below 4500; and children living in low traffic (LT) areas, defined as at least 300 m from any road with over 25,000 AADT (Figure 1).

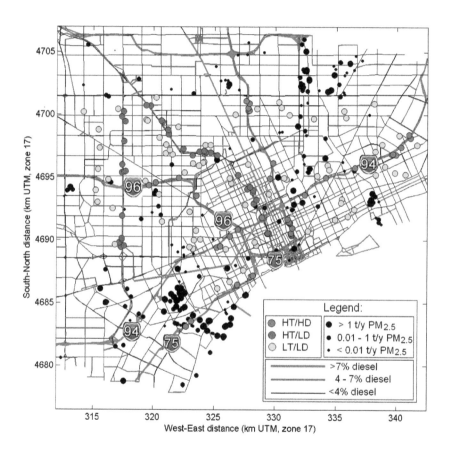

FIGURE 1: Modeling domain for the NEXUS study. Major highways are shown as red and blue lines (for >7% diesel and 4%–7% diesel fraction) and other roads–as black lines. Model receptors are shown in red, blue and green circles for the HD, LD and LT traffic exposure group, respectively. Stationary sources are shown as black dots (symbol size indicates the magnitude of $PM_{2.5}$ annual emissions).

We first estimated pollutant-specific local-scale air concentrations for stationary and area sources using AERMOD. This model utilized information on local emission sources for these two sectors and local meteorological conditions to estimate hourly average concentrations at multiple receptors in each of the three exposure groups. Emission data for major stationary sources and airport sources were obtained from the NEI. For mobile sources, we used a recently developed line source dispersion model RLINE [24,25]. Roadway emissions were estimated using detailed road network locations and a bottom-up methodology for roadway emissions [20], and further elaborated in [28].

An analysis of wind patterns for the year 2010 based on hourly meteorological observations from the NWS stations within and around the study area (Detroit City airport, Detroit Metro airport, Windsor airport) determined that the Detroit-City airport station was most representative of the NEXUS modeling domain, and which also had the most data completeness objective. Hourly surface observations from Detroit City, in combination with data from the nearest upper air station (DTX-72632 Oakland County) were used for the simulation period to drive the modeling. The land characteristics around the station were determined and the AERSURFACE model was applied. The AERMET program was used to process the meteorological data from the Detroit City airport and DTX upper air station for input into AERMOD.

Emissions within the 30 × 40 km source region centered on the NEXUS participants in Detroit were extracted from the NEI 2008 by major source categories (area, point, onroad and off-road mobile) for the pollutants of interest. Sources located in Macomb, Oakland, and Wayne counties in Michigan, and Essex County in Ontario, Canada were included. Area sources such as port- and airport-type sources in the study area were also included.

For stationary point sources, the location, emission rate, and individual stack parameters (e.g., stack height, exit velocity) were used. Other non-stack emissions (such as smaller sources with no stack parameters, fugitive emissions, and emissions from nonroad mobile sources) were modeled as area sources. County-level NEI area source emissions were spatially re-allocated to 1 km × 1 km grid-cell resolution using spatial surrogates within the SMOKE emissions processor [29]. Airport area sources with a polygon-shaped area corresponding to their actual locations were used as an input to the model. Stationary sources were temporally allocated using SMOKE. The SMOKE processor contains monthly, weekly, diurnal-weekday and diurnal-weekend profiles. A seasonal profile was calculated from the monthly profiles. The final temporal allocation yields an emission rate for each hour of the weekday/Saturday/Sunday for the entire year.

For onroad mobile source emissions, the methodology described in [20] is followed that produces a spatially and temporally resolved mobile source emissions inventory (i.e., hourly emissions for all pollutants modeled, by vehicle class and road link). This methodology was successfully applied in previous studies for New Haven, Atlanta and Baltimore [16,22,30]. In this study, detailed information including the geometry of the road network, traffic volumes, temporal allocation factors, fleet mixes and pollutant-specific emission factors, assembled from a variety of sources, were used in combination with meteorological inputs to generate link-based emissions for use in dispersion modeling to estimate pollutant concentrations due to traffic [28]. The total emissions were calculated from emission factors multiplied by traffic activity for each road link to provide inputs for RLINE model simulations across the NEXUS study domain for a 1.5 years period (Fall 2010–Spring 2012). In order to evaluate the differences in near-road pollutant gradients between the three selected traffic exposure groups (low diesel LD, high diesel HD and low traffic LT), the receptor grids were refined within each NEXUS sub-area (including the participants homes and schools). A mini-grid of receptors was placed near each NEXUS participant's home and school consisting of a rectangular receptor grid on 50 m centers as indicated in Figure 2. Depending on the number of receptors used, mini-grids gave anonymity to 50 or 100 m, a distance sufficient to protect the participants' identity.

Exposure metrics were calculated from mini-grids to produce estimates for each NEXUS location. For NEXUS locations in the near road group, there are 85 near-road grids. The near-road grids contain 24 modeled receptors, and a weighted interpolation between modeled grid rows was performed based on the actual distance between the participant's home and the nearest major roadway to estimate the hourly concentration. Other locations were modeled with 5-point receptor grids (using five receptors on and around the home) and the hourly concentration was estimated by taking an average of the modeled concentrations at the five points.

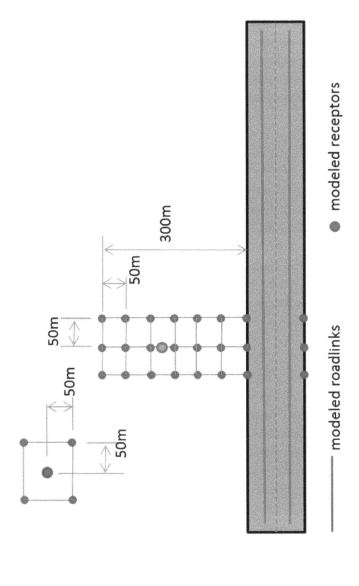

FIGURE 2: Model receptors near roadways: 24-receptor mini-grid network.

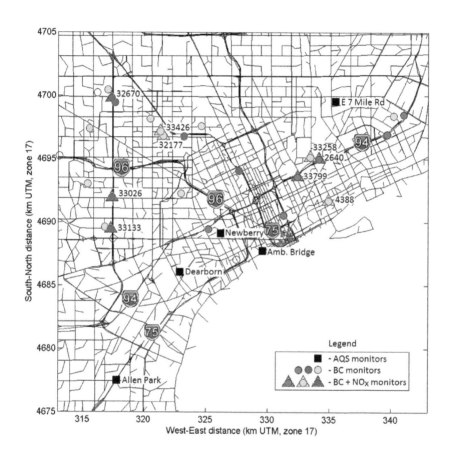

FIGURE 3. Locations of PM2.5, black carbon (BC) and NOx monitors at NEXUS (●, ▲) and AQS sites (■). (Notes: Colors of symbols denote roadway classification as described in Figure 1; numbers next to the NEXUS site locations indicate measurement site ID).

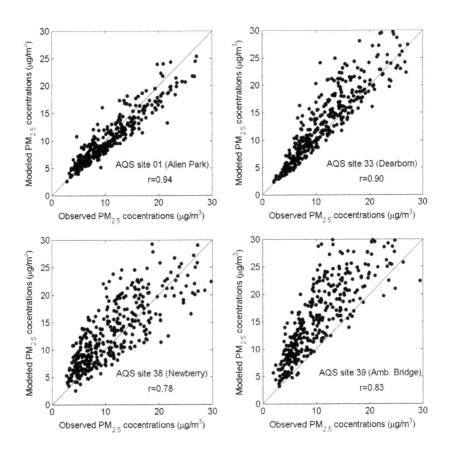

FIGURE 4: Model to monitor comparison: daily average PM2.5 concentrations for one-year period of 2010 at four AQS sites in the Detroit modeling domain.

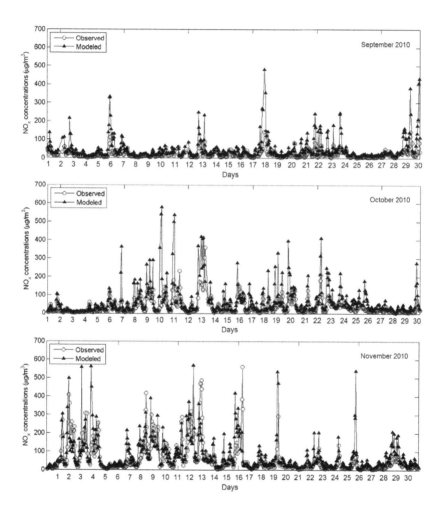

FIGURE 5: Model to monitor comparison: time series of hourly NOx concentrations at the AQS site 26-163-0019 (E. 7 Mile Road) for three-month period September-November 2010.

From hourly concentration, exposure metrics were calculated for the following time periods: 24 h (daily); 1–6 (a.m. off-peak); 7–8 (a.m. peak); 9–14 (mid-day); 15–17 (p.m. peak); and 18–24 (p.m. off-peak). These hours correspond to the reported local-time (e.g., hour 1 represents from 12:01 a.m.–1:00 a.m.). These are calculated with a 70% completeness criterion for the hourly meteorology in each time period. These daily exposure metrics for CO, NOx, $PM_{2.5}$ and its components (EC and OC), capturing spatial and temporal variability across health study domain (Fall 2010–Spring 2012) were used in the epidemiologic analyses.

11.3 RESULTS AND DISCUSSION

Model results were compared to ambient monitoring data in Detroit. There are two sets of monitoring data for model evaluation: (1) from the routine observational network (AQS); and (2) from the intensive monitoring campaign which was part of the NEXUS study. There are five AQS monitoring stations in the modeling domain: four $PM_{2.5}$ monitors (Allen Park, Dearborn, Newberry School, Ambassador Bridge) and one NOx monitor (East 7 mile road), as indicated in Figure 3. A comparison between modeled daily average $PM_{2.5}$ concentrations for one-year period of 2010 at observed $PM_{2.5}$ concentrations at all four AQS sites is shown in Figure 4. Model results correlate well with observed data (r ranges from 0.78 to 0.94) and are generally within a factor of two from observations. The Allen Park site near I-75 and southwest of stationary sources has best comparison vs. other sites closer to large sources. There is more scatter at the "Newberry" and "Ambassador Bridge" sites, likely due to uncertainties in spatial allocation of emissions near these locations. These sites are impacted by local emission sources modeled as 1 km × 1 km area sources in AERMOD. In contrast, the "Dearborn" site is impacted by industrial sources modeled as stacks with their known locations. For NOx, only one monitoring site was available in the modeling domain. The "East 7 mile" site is in the North-Eastern corner of the modeling domain, away from major highways. Figure 5 compares time series of modeled and observed hourly NOx concentrations at the "East 7 mile" site for September–November 2010. Modeled concentrations generally follow the time series of observed data, however there are some over-predictions at certain hours likely due to uncertainties in emissions from traffic. The monitoring site is away from major highways, therefore the observed concentrations are influenced by emissions from local roads and regional sources. Unlike major highways, estimating emissions from local roads is more challenging because of uncertainties in road locations, traffic activity and fleet distribution. The results of statistical analyses (i.e., Mean Bias, Mean Error, R, FAC2) comparing the modeled and measurement data from five AQS monitoring stations in the modeling domain are summarized in Table 1.

TABLE 1: Statistics metrics for the model-to-monitor comparison at the five AQS monitoring stations for $PM_{2.5}$ and NOx.

Pollutant	PM2.5				NOx
Site	261630001	261630033	261630038	261630039	261630019
Obs. Mean	10.865	11.694	11.050	11.619	32.656
Model Mean	10.370	13.646	14.233	18.243	62.255
Mean Bias	−0.495	1.952	3.183	6.624	29.598
Mean Error	2.420	4.254	5.798	7.834	35.654
R	0.760	0.624	0.480	0.502	0.515
FAC2	0.965	0.905	0.818	0.787	0.616
Pairs	8365	8438	8297	8455	8100

The modeling provides an opportunity to compare the relative contributions of various sources: stationary sources (i.e., AERMOD), roadways (i.e., RLINE), urban background (i.e., STOK), and total (Hybrid). Figure 6 compares distributions of modeled and observed concentrations for $PM_{2.5}$ (all four AQS sites combined) and NOx (one AQS site) for 2010, and also shows relative contributions of various sources. As can be seen from Figure 6, the relative contribution of roadways is very small for $PM_{2.5}$ but quite high for NOx, whereas urban background is more significant for $PM_{2.5}$ than for NOx. The difference in relative contribution of roadway emissions to the total concentration between pollutants is further illustrated in Figure 7 using a single receptor site near the I-94 freeway as an example. The model predicts steep gradients of near-road concentrations for all pollutants (CO, NOx and PM$_{2.5}$) at the modeled receptor site near I-94. However, the background contribution is different for these pollutants. For CO, the roadway contribution is high within 100 m from the roadway, but after 100 m it diminishes to levels below the background. For NOx, the background is low and roadway impact dominates at this site. For PM2.5, the background dominates and primary impact of roadway emissions contributes only about 10%–25% of the total concentration.

Measurements of air pollutant exposures also have uncertainties, such as from the measurement method or instrument, as well as whether the measurement captures actual air pollutant exposures or is a surrogate for it (e.g., central site monitors). Although the sub-daily modeled exposure metrics may have greater uncertainty than daily or longer-term averages, few monitoring methods exist that can measure exposures with time resolution below daily averages. Collecting limited high-time resolution measurements for comparison with model predictions is one approach to help identify potential contributors to the modeling uncertainty. In addition to observational data from the routine monitoring network, we also used monitoring data from the 2010 intensive monitoring campaign of the NEXUS study. During the September-November 2010 study period, black carbon (BC) measurements were made at 25 NEXUS home locations and NOx was measured at nine NEXUS homes (Figure 3). Figure 8 compares modeled and observed concentrations at selected NEXUS homes for NOx and BC. As can be seen from the figure, the model generally captures the time series of observed NOx concentrations. However, at some sites and for some specific hours, the model under-predicts concentrations (e.g., at site ID = 33,133 or ID = 32,177, 6–8 a.m. on 29 September 2010) or over-predicts (e.g., at site ID = 33,426, 6–8 a.m. on 29 September 2010) concentrations at some locations. This discrepancy can be explained by the uncertainty in hourly traffic activity at the road link level. Typically, time-resolved traffic information at a link level is not available and sophisticated algorithms are used to estimate such traffic emissions for individual road links. Nevertheless, except for some events, the model can capture the magnitude and time patterns of near road pollutant concentrations, critical for the exposure and health studies. For BC, the model performance was similar to NOx, if not better at the sites shown.

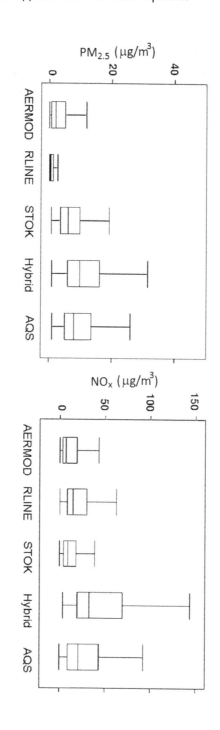

FIGURE 6: Distributions of modeled and observed PM$_{2.5}$ and NOx concentrations for 2010 at the AQS monitoring sites. (all four PM$_{2.5}$ averaged, and one NOx site).

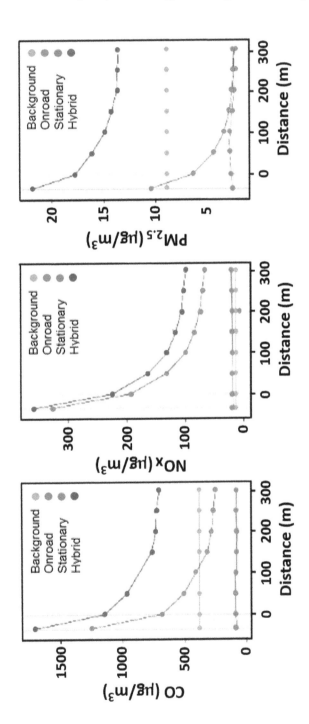

FIGURE 7: Near-road pollutant gradients of CO, NOx and PM$_{2.5}$ concentrations (2010 annual average) from a mini-grid of 24 model receptors near the I-94 freeway.

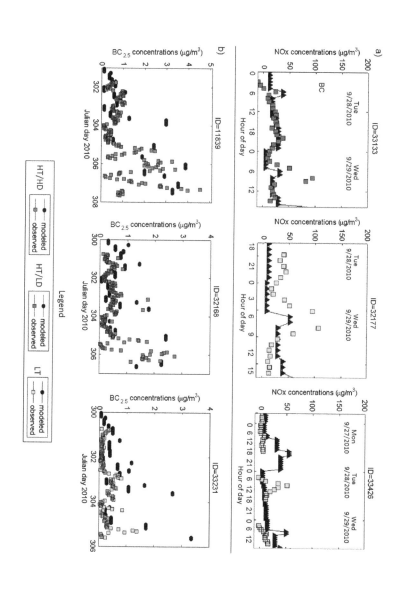

FIGURE 8: Comparison of modeled exposure metrics and observed concentrations for NOx at six different NEXUS monitoring sites.

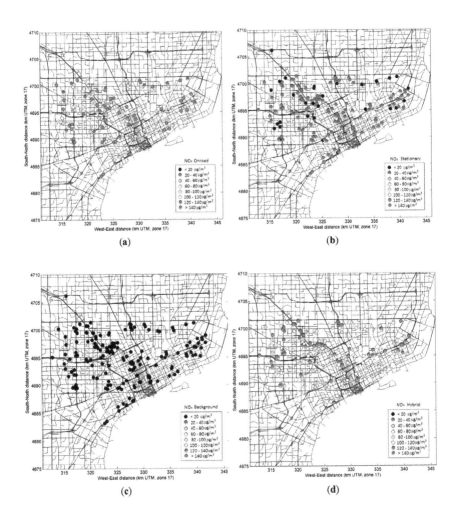

FIGURE 9: Spatial maps of modeled daily NOx concentrations averaged during September-October 2010, showing contributions from mobile sources (a); stationary sources (b); urban background (c); and total (d).

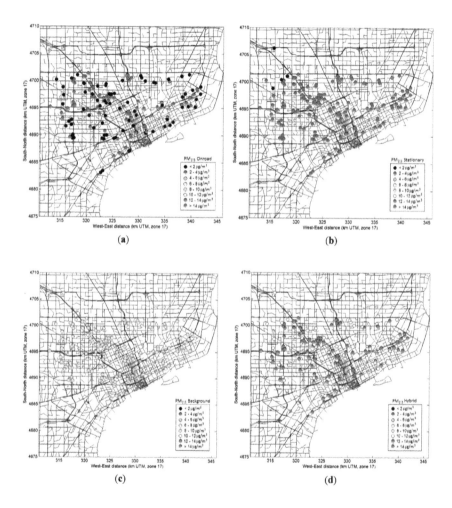

FIGURE 10: Spatial maps of modeled daily PM$_{2.5}$ concentrations averaged during September–October 2010, showing contributions from mobile sources (a); stationary sources (b); urban background (c); and total (d).

The model-based exposure metrics for CO, NOx, $PM_{2.5}$ and its components (EC and OC), were calculated from hourly predictions and were able to capture the spatial and temporal variability across the health study domain. The modeling approach also allowed estimating relative contributions of roadways vs. stationary sources and urban background. Figure 9 and Figure 10 show spatial maps of modeled daily NOx and $PM_{2.5}$ concentrations averaged over the study period (September–October 2010) and the relative contributions of mobile sources, stationary sources, and urban background as well as the total (hybrid). For both NOx and $PM_{2.5}$, the urban background was nearly uniform across the domain, while mobile source contributions varied across the domain—with higher concentrations next to major roadways and lower concentrations away from roads. The overall mobile source contribution, however, was not the same for NOx and $PM_{2.5}$. For NOx, urban background contributes less than half of total concentrations, whereas for $PM_{2.5}$, the urban background dominated and the local impact of mobile sources was less than 30%. Also, stationary source contributions for $PM_{2.5}$ were of similar range to mobile sources.

11.4 CONCLUSIONS

Here we presented an application of a hybrid modeling approach to estimate exposure metrics in support of an urban scale epidemiologic study of exposures to traffic-related pollutants for children with asthma living near major roadways in Detroit, Michigan. The modeling approach involved the development and use of a detailed emissions inventory and multiple dispersion models to estimate ambient air pollution concentrations. The emissions inventory was based on a detailed geometry of the road network, traffic volumes, temporal allocation factors, fleet mix, and pollutant specific emission factors. These road-link emissions were used as inputs to RLINE, the newly developed dispersion model specifically designed for near-road applications. Thus, the model-based exposure metrics provided the temporal and spatial resolution needed for the epidemiologic study. Using a novel mini-grid approach, the modeling was able to resolve near-road air pollutant gradients. The hybrid modeling approach also provided an opportunity to compare relative contributions of various sources: stationary sources, roadways, urban background, and total. While near-road gradients of roadway emissions within 300 m were strong for all pollutants, their relative contributions to the total concentration varied by pollutant.

The hybrid modeling approach used in NEXUS provides new information regarding exposure to traffic-related air pollutants that is not captured by simpler exposure metrics (such as traffic intensity and distance to roads) commonly used in environmental epidemiology studies of traffic-related air pollution. Such additional information on strong spatial and temporal variation of pollutant concentrations and the relative contribution of various source categories to the total concentration could benefit future traffic-related health assessments. The hybrid modeling approach used in NEXUS could be also used for estimating exposures in other epidemiological studies where adequate measurements of traffic- or other source-related air pollutants are not feasible.

REFERENCES

1. Health Effects Institute (HEI). Traffic-Related Air Pollution: A Critical Review of the Literature on Emissions, Exposure, and Health Effect; HEI: Boston, MA, USA, 2010.
2. Cakmak, S.; Mahmud, M.; Grgicak-Mannion, A.; Dales, R. The influence of neighborhood traffic density on the respiratory health of elementary schoolchildren. Environ. Int. 2012, 39, 128–132.
3. Rosenbloom, J.; Wilker, E.; Mukamal, K.; Schwartz, J.; Mittleman, M. Residential proximity to major roadway and 10-year all-cause mortality after myocardial infarction. Circulation 2012, 125, 2197–2203.
4. Chen, H.; Goldberg, M.; Burnett, R.; Jerrett, M.; Wheeler, A.; Villeneuve, P. Long-Term exposure to traffic-related air pollution and cardiovascular mortality. Epidemiology 2013, 24, 35–43.
5. Gehring, U.; Gruzieva, O.; Agius, R.M.; Beelen, R.; Custovic, A.; Cyrys, J.; Eeftens, M.; Flexeder, C.; Fuertes, E.; Heinrich, J.; et al. Air pollution exposure and lung function in children: The ESCAPE Project. Environ. Health Perspect. 2013, 121, 1357–1364.
6. Miranda, M.; Edwards, S.; Chang, H.; Auten, R. Proximity to roadways and pregnancy outcomes. J. Expos. Sci. Environ. Epidemiol. 2013, 23, 32–38.
7. Batterman, S.; Burke, J.; Isakov, V.; Lewis, T.; Mukherjee, B.; Robins, T. A comparison of exposure metrics for traffic-related air pollutants: application to epidemiology studies in detroit, michigan. Int. J. Environ. Res. Public Health. submitted.

8. Wheeler, A.; Smith-Doiron, M.; Xu, X.; Gilbert, N.; Brook, J. Intra-Urban variability of air pollution in Windsor, Ontario-measurement and modeling for human exposure assessment. Environ. Res. 2008, 106, 7–16.
9. Matte, T.D.; Ross, Z.; Kheirbek, I.; Eisl, H.; Johnson, S.; Gorczynski, J.E.; Kass, D.; Markowitz, S.; Pezeshki, G.; Clougherty, J.E. Monitoring intraurban spatial patterns of multiple combustion air pollutants in New York City: Design and implementation. J. Expos. Sci. Environ. Epidemiol. 2013, 23, 223–231.
10. Karner, A.; Eisinger, D.; Niemeier, D. Near-Roadway air quality: Synthesizing the findings from real-world data. Environ. Sci. Technol. 2010, 44, 5334–5344.
11. Beckx, C.; Panis, L.I.; Uljee, I.; Arentze, T.; Janssens, D.; Wets, G. Disaggregation of nation-wide dynamic population exposure estimates in The Netherlands: Applications of activity-based transport models. Atmos. Environ. 2009, 43, 5454–5462.
12. Hatzopoulou, M.; Miller, E.J. Linking an activity-based travel demand model with traffic emission and dispersion models: Transport's contribution to air pollution in Toronto. Transp. Res. Part D Transp. Environ. 2010, 15, 315–325.
13. McConnell, R.; Islam, T.; Shankardass, K.; Jerrett, M.; Lurmann, F.; Gilliland, F.; Gauderman, J.; Avol, E.; Kunzli, N.; Yao, L.; et al. Childhood incident asthma and traffic-related air pollution at home and school. Environ. Health Persp. 2010, 118, 1021–1026.
14. Gruzieva, O.; Bellander, T.; Eneroth, K.; Kull, I.; Melén, E.; Nordling, E.; van Hage, M.; Wicjman, M.; Moskalenko, V.; Hulchiy, O.; et al. Traffic-Related air pollution and development of allergic sensitization in children during the first 8 years of life. J. Allergy Clin. Immunol. 2013, 129, 240–246.
15. Sørensen, M.; Hoffmann, B.; Hvidberg, M.; Ketzel, M.; Jensen, S.S.; Andersen, Z.J.; Tjønneland, A.; Overvad, K.; Raaschou-Nielse, O. Long-term exposure to traffic-related air pollution associated with blood pressure and self-reported hypertension in a Danish cohort. Environ. Health Persp. 2012, 120, 418–424.
16. Sarnat, S.E.; Sarnat, J.A.; Mulholland, J.; Isakov, V.; Ozkaynak, H.; Chang, H.H.; Klein, M.; Tolbert, P.E. Application of alternative spatiotemporal metrics of ambient air pollution exposure in a time-series epidemiological study in Atlanta. J. Expos. Sci. Environ. Epidemiol. 2013, 23, 593–605.
17. Gurram, S.; Stuart, A.L.; Pinjari, A.R. Impact of travel activity and urbanicity on exposures to ambient nitrogen oxides and on exposure disparities between sub-populations in Tampa, Florida. Air Qual. Atmos. Health 2014.
18. Byun, D.; Schere, K. Review of the governing equations, computational algorithms, and other components of the models-3 Community Multiscale Air Quality (CMAQ) modeling system. Appl. Mech. Rev. 2006, 59, 51–77.
19. Cimorelli, A.J.; Perry, S.G.; Venkatram, A.; Weil, J.C.; Paine, R.J.; Wilson, R.B.; Lee, R.F.; Peters, W.D.; Brode, R.W. AERMOD: A dispersion model for industrial source applications. Part I: General model formulation and boundary layer characterization. J. Appl. Meteorol. Climatol. 2005, 44, 682–693.
20. Cook, R.; Isakov, V.; Touma, J.S.; Benjey, W.; Thurman, J.; Kinnee, E.; Ensley, D. Resolving local-scale emissions for modeling air quality near roadways. J. Air Waste Manag. Assoc. 2008, 58, 451–61.
21. Dionisio, K.L.; Isakov, V.; Baxter, L.K.; Sarnat, J.A.; Sarnat, S.E.; Burke, J.; Rosenbaum, A.; Graham, S.E.; Cook, R.; Mulholland, J.; et al. Development and evalua-

tion of alternative approaches for exposure assessment of multiple air pollutants in Atlanta, Georgia. J. Expos Sci. Environ. Epidemiol. 2013, 23, 581–592.

22. Isakov, V.; Touma, J.; Burke, J.; Lobdell, D.; Palma, T.; Rosenbaum, A.; Kozkaynak, H. Combining regional-and local-scale air quality models with exposure models for use in environmental health studies. J. Air Waste Manag. Assoc. 2009, 59, 461–472.

23. Vette, A.; Burke, J.; Norris, G.; Landis, M.; Batterman, S.; Breen, M.; Isakov, V.; Lewis, T.; Gilmour, M.I.; Kamal, A.; et al. The Near-Road Exposures and Effects of Urban Air Pollutants Study (NEXUS): Study design and methods. Sci. Total Environ. 2013, 448, 38–47.

24. Snyder, M.G.; Venkatram, A.; Heist, D.K.; Perry, S.G.; Petersen, W.B.; Isakov, V. RLINE: A line source dispersion model for near-surface releases. Atmos. Environ. 2013, 77, 748–756.

25. Venkatram, A.; Snyder, M.G.; Heist, D.K.; Perry, S.G.; Petersen, W.B.; Isakov, V. Re-formulation of plume spread for near-surface dispersion. Atmos. Environ. 2013, 77, 846–855.

26. U.S. Environmental Protection Agency. The 2008 National Emissions Inventory. Available online: http://www.epa.gov/ttn/chief/net/2008inventory.html (accessed on 12 February 2013).

27. Arunachalam, S.; Valencia, A.; Akita, Y.; Serre, M.; Omary, M.; Garcia, V.; Isakov, V. Estimating regional background air quality using space/time ordinary kriging to support exposure studies. Int. J. Environ. Res. Public Health. submitted.

28. Snyder, M.G.; Arunachalam, S.; Isakov, V.; Talgo, K.; Naess, B.; Valencia, A.; Davis, N.; Cook, R. Creating mobile source emissions for an urban-scale air quality assessment to support exposure studies. Int. J. Environ. Res. Public Health. submitted.

29. Houyoux, M.R.; Vukovich, J.M.; Coats, C.J., Jr.; Wheeler, N.J.M. Emission inventory development and processing for the Seasonal Model for Regional Air Quality (SMRAQ) project. J. Geophys. Res. 2000, 105, 9079–9090.

30. Lobdell, D.T.; Isakov, V.; Baxter, L.; Touma, J.S.; Smuts, M.B.; Özkaynak, H. Feasibility of assessing public health impacts of air pollution reduction programs on a local scale: New Haven case study. Environ. Health Persp. 2011, 119, 487–493.

PART V

AGRICULTURE

CHAPTER 12

Characteristics and Emission Budget of Carbonaceous Species from Post-Harvest Agricultural-Waste Burning in Source Region of the Indo-Gangetic Plain

PRASHANT RAJPUT, MANMOHAN SARIN, DEEPTI SHARMA, AND DARSHAN SINGH

12.1 INTRODUCTION

The hot-spots of atmospheric pollutants over the Indo-Gangetic Plain (IGP) and a thick layer of haze advecting towards the Bay of Bengal (BoB) during the wintertime have been documented through MODIS (MODerate resolution Imaging Spectroradiometer) imageries (Ramanathan et al., 2007). This has led to the suggestion on the impact of aerosols (particularly black carbon) and warming trend (heating rate: 0.15–0.30 K/decade) in Northern India (Ramanathan et al., 2007). The IGP occupies ~15% of the geographical area in south Asia, holds about 42% of the total population and accounts for 45% of food production (Gupta et al., 2004; Badarinath

*Characteristics and Emission Budget of Carbonaceous Species from Post-Harvest Agricultural-Waste Burning in Source Region of the Indo-Gangetic Plain. © Rajput P, Sarin M, Sharma D, and Singh D. Current Status and Prospects of Biodiesel Production from Microalgae. Tellus B **66** (2014), http:// dx.doi.org/10.3402/tellusb.v66.21026. Licensed under Creative Commons Attribution 4.0 International License, http://creativecommons.org/licenses/by/4.0/.*

et al., 2006). The two major crops of paddy (rice) and wheat grown in the IGP contribute nearly 85% of the entire production in south Asia. On a regional scale, 90–95% of rice and wheat crop rotation in India [total area under rice-wheat cultivation is ~20 million hectares (m ha)] are located in Punjab, Haryana and western part of Uttar Pradesh in the IGP. The burning of agricultural-waste in open fields for crop rotation is a common practice on annual and seasonal basis in north-west region of the IGP (Gupta et al., 2004; Badarinath et al., 2006; Punia et al., 2008).

The large-scale biomass burning emission in the IGP addresses the issue of contribution from secondary organic aerosol (SOA) and potential loss of atmospheric chemical constituents (O_3, NOx and OH radical) via chemical reactions (Rengarajan et al., 2007; Rajput et al., 2011b; Ram and Sarin, 2011). During the wintertime (December–February), emissions from bio-fuel burning and fossil-fuel combustion sources are trapped in the lower atmosphere due to shallow planetary boundary layer associated with fog-haze conditions in the IGP (Rengarajan et al., 2007; Rajput et al., 2011b; Ram et al., 2012). It is relevant to state that the post-harvest paddy-residue burning emissions can have a significant impact during the wintertime on the aerosol composition over Northern India. In this context, we present extensive data set on the characterisation of two major crop-residue (paddy- and wheat-residue) burning emissions in the IGP. The emission budget of organic and elemental carbon (OC, EC) and polycyclic aromatic hydrocarbon contributes significantly to the global emission scenario from agricultural-waste burning. Furthermore, we have also assessed the spatial variability in OC/EC ratio (as high as 6–10) as a characteristic feature of carbonaceous aerosols from different geographical locations during the wintertime in the IGP.

12.2 METHODOLOGY

12.2.1 SAMPLING SITE DESCRIPTION

The large-scale biomass burning emission in Northern India (mainly in Punjab, Haryana and western Uttar Pradesh; $2°×2°=48400$ sq km; Fig. 1), extending from north to the north-western part of the IGP, is a common practice followed by farmers on annual and seasonal basis. The post-harvest paddy-residue burning during October–November and wheat-residue burning during April–May are conspicuous features in the IGP. The intermediate period from December to February (wintertime) experiences fog formation, a manifestation of shallower planetary boundary layer, emissions from bio-fuel burning and fossil-fuel combustion sources and moisture from western disturbances. In order to assess the chemical characteristics of ambient aerosols from these emission sources, samples of $PM_{2.5}$ (particulate matter with aerodynamic diameter $\leq_{2.5}$ μm) were collected from a sampling site at Patiala [30.2°N, 76.3°E; 250 m above mean sea level (amsl)], where nearly 84% of the land area is under cultivation (Badarinath et al., 2006). During the period of SW-monsoon (June–September), frequent wet precipitation events amounting to 80–90% of the total annual precipitation (~1000 mm) wash out ambient aerosols. After monsoon, aerosol composition at the sampling site (upwind of the major population and industrial pollution sources) is considered to be representative of emissions from open paddy- (October–November) and wheat-residue burning (April–May) (Rajput et al., 2011b).

12.2.2 ANALYTICAL METHODS

Aerosol samples were collected using a high-volume sampler (flow rate: 1.2 m³ minutes⁻¹) by filtering ambient air through the pre-combusted (at 350°C for ~6 h) tissuquartz filters (PALLFLEX™, 2500QAT-UP, 20 cm×25 cm). A total of 59 $PM_{2.5}$ samples were collected during paddy-residue burning emissions (October–November in 2008 and 2010), of which initial 33 samples were integrated for ~24 h each (in the first-campaign in 2008) and the remaining 26 samples for 8–10 h each (daytime in the second-campaign in 2010), to collect the adequate aerosol mass on tissuquartz filters. However, during wheat-residue burning emissions (April–May in 2009 and 2011), all samples (n=31) were collected during the daytime, integrating each sample for 8–10 h and so was the case during wintertime (n=51; December 8–March 9 and December 10–March 11).

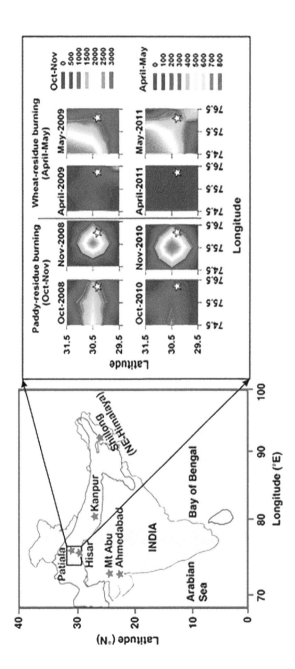

FIGURE 1: Map of study locations: Patiala, Hisar and Kanpur in the Indo-Gangetic Plain (shown as yellow shaded area), at Barapani near Shillong in NE-Himalaya and at Ahmedabad and Mt Abu in semi-arid western India. MODIS derived fire-counts (plotted on the right) during the paddy- and wheat-residue burning period in the source region; sampling site at Patiala (shown by open star) is located downwind of major field-fires.

TABLE 1: Concentrations of carbonaceous species and characteristic ratios (Av±sd given in parenthesis) from biomass burning emissions and fossil-fuel combustion in the Indo-Gangetic Plain (IGP)

Aerosol species	Paddy-residue burning (Oct–Nov; n=59)	Wheat-residue burning (April–May; n=31)	Difference[a] (two-tailed t-test)	Fossil-fuel and bio-fuel emissions (Dec–March; n=51)
$PM_{2.5}$ ($\mu g\ m^{-3}$)	60–391	18–123		19–244
	(195±87)	(50±23)		(124±58)
$OC/PM_{2.5}$ (%)	21–50	19–36	S (t=4.9)	12–31
	(33±7)	(26±5)		(22±4)
$EC/PM_{2.5}$ (%)	2–6	4–12	S (t=9.0)	2–9
	(3.5±1.1)	(6.9±$_{2.5}$)		(4.6±1.8)
OC/EC[b]	4–26	2.0–6.5	S (t=26)	1.9–10.1
	(10.6±1.6)	(3.0±0.4)		(4.2±0.8)
EC/TC[b]	0.04–0.18	0.13–0.33	S (t=58)	0.09–0.35
	(0.04±0.00)	(0.19±0.02)		(0.15±0.01)
WSOC/OCb	0.41–0.91	0.43–0.79	S (t=15)	0.45–0.86
	(0.52±0.02)	(0.60±0.03)		(0.62±0.03)
nss-K+/OC[b]	0.03–0.16	0.04–0.15	S (t=62)	0.01–0.13
	(0.06±0.01)	(0.14±0.01)		(0.06±0.01)
ΣPAHs ($ng\ m^{-3}$)	3.2–59.1	1.2–17.0		2.1–47.9
	(27.1±16.7)	(4.1±3.5)		(16.9±10.4)
ΣPAHs/OC ($mg\ g^{-1}$)[b]	0.07–0.96	0.22–1.22	S (t=5.5)	0.12–0.82
	(0.37±0.04)	(0.30±0.08)		(0.65±0.07)
ΣPAHs/EC ($mg\ g^{-1}$)[b]	0.80–10.19	0.69–2.35	S (t=22.3)	1.32–9.58
	(4.25±0.72)	(1.30±0.20)		(3.41±0.66)
PAHs isomer ratios				
3-ring ANTH/(ANTH+PHEN)	(0.19±0.08)	(0.11±0.04)	S (t=5.2)	(0.15±0.05)
4-ring FLA/(FLA+PYR)	(0.46±0.02)	(0.50±0.03)	S (t=7.5)	(0.48±0.02)
BaA/(BaA+CHRY+TRIPH)	(0.25±0.05)	(0.24±0.09)	IS (t=0.6)	(0.28±0.04)
5-ring BaP/(BaP+B[b,j,k]FLA)	(0.25±0.05)	(0.26±0.06)	IS (t=0.8)	(0.27±0.05)
6-ring IcdP/(IcdP+BghiP)	(0.49±0.03)	(0.45±0.05)	S (t=4.7)	(0.50±0.05)

[a]*Comparison between paddy- and wheat-residue burning emissions.* [b]*Correlation analysis. S (significant difference for p<0.05) and IS (insignificant difference for p>0.05). ΣPAHs (PM$_{2.5}$-bound) include: NAPH, naphthalene; ACY, acenaphthylene; 2-BrNAPH, 2-bromonaphthalene; ACE, acenaphthene; FLU, fluorene; PHEN, phenanthrene; ANTH, anthracene; FLA, fluoranthene; PYR, pyrene; BaA, benzo[a]anthracene; CHRY+TRIPH, chrysene/triphenylene; B[b,j,k]FLA, benzo[b+j+k]fluoranthene; BaP, benzo[a]pyrene; IcdP, indeno[1,2,3-cd]pyrene; D[ah,ac]ANTH, dibenzo[a,h+a,c]anthracene}; BghiP, benzo[g,h,i]perylene.*

The meteorological conditions (temperature, relative humidity and wind) are nearly same from year-to-year during the paddy-residue burning emissions (October–November, 2008 and 2010), and so is the case during wheat-residue burning emissions (April–May, 2009 and 2011). The meteorological parameters (temperature, relative humidity and wind) during wintertime also do not exhibit inter-annual variability for the data set discussed in this study. The elemental carbon (EC) and organic carbon (OC) were measured on thermo-optical carbon analyser (Sunset Laboratory) using the NIOSH (National Institute for Occupational Safety and Health) protocol (Birch and Cary, 1996), water-soluble organic carbon (WSOC) on total OC analyser (Shimadzu; TOC 5000 A), water-soluble potassium on ion-chromatograph (Dionex®) and particulate-bound polycyclic aromatic hydrocarbons (here after referred as PAHs and in Table 1) on gas chromatograph coupled with a mass spectrometer (GC-MS, Agilent: 7890A/5975C) (Rengarajan et al., 2007; Rajput et al., 2011a, 2011b, 2013; Ram et al., 2012).

12.3 RESULTS AND DISCUSSION

12.3.1 AEROSOL CHARACTERISTICS FROM POST-HARVEST BIOMASS BURNING EMISSIONS IN THE IGP

Post-harvest paddy-residue (with moisture content: 40–50%) is burnt in open fields (under ambient atmospheric conditions of 24–30°C temperature and 51–65% relative humidity), during October–November in the IGP (Gupta et al., 2004; Badarinath et al., 2006; Punia et al., 2008). The post-harvest wheat-residue (with moisture content: <5%) is burnt under ambient temperature ranging from 33 to 38°C at low relative humidity (34–41%), during April–May. The $PM_{2.5}$ mass concentration averages around 200 µg m^{-3} during the paddy-residue burning emission (Table 1; Fig. 2) and ~50 µg m^{-3} during the wheat-residue burning emission. The average contributions of OC, EC and nss-K$^+$ to $PM_{2.5}$ from paddy-residue burning emission are 33, 3.5 and 2.4%, respectively (Fig. 2). From wheat-residue burning emission, the average contributions of OC, EC and nss-K$^+$ to $PM_{2.5}$ are 26, 7 and 2.4%, respectively.

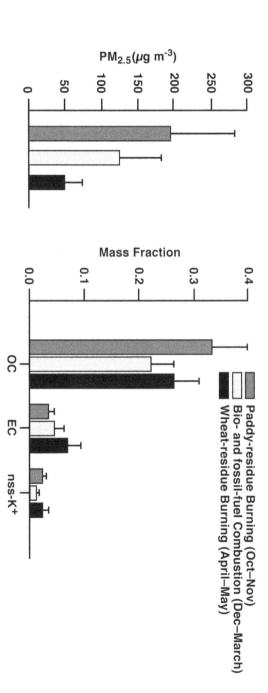

FIGURE 2: Temporal variability in PM2.5 and mass fractions of OC, EC and nss-K⁺ associated with different emissions from source region (Patiala, Fig. 1) in the Indo-Gangetic Plain. The nss-K⁺=K⁺aerosol−0.037*Na+aerosol; where K⁺/Na⁺ mass ratio of 0.037 is used for sea-salt contribution of K⁺ (Keene et al., 1986).

The ΣPAHs (sum of 16 PAHs: 2- to 6-ring, reference is made to Table 1) average concentration from paddy-residue burning is 27.1 ng m^{-3} and from wheat-residue burning emission is 4.1 ng m^{-3}. The contribution of 4- to 6-ring PAHs (Σ(4-to6-)PAHs) to the total PAHs (ΣPAHs) is 95±2% for the two biomass burning emission sources. Further differences in the two biomass burning emission sources are assessed by the relative distribution of ring-wise PAHs with respect to the 6-ring PAHs. The 4-/6-ring and 5-/6-ring distributions also exhibit significant differences between the two biomass sources: relatively high ratios are recorded for the paddy-residue burning emission. The ΣPAHs/OC ratio of 0.4 mg g^{-1} from paddy-residue burning is relatively high (Table 1) than that from the wheat-residue burning emission (0.30 mg g^{-1}). Furthermore, the ΣPAHs/EC ratio of 4.3 mg g^{-1} from paddy-residue burning is also higher (Table 1) than that from the wheat-residue burning emission (1.30 mg g^{-1}). The combustion of moist paddy-residue (with moisture content: 40–50%) during open biomass burning in October–November is responsible for higher contribution of OC to PM$_{2.5}$ and high-molecular weight PAHs (4- to 6-rings) in OC. In contrast, under dry weather conditions, the emissions from wheat-residue burning (with moisture content<5%) gives higher contribution of EC in PM$_{2.5}$ as compared to that from paddy-residue burning emission (EC/PM$_{2.5}$: 3.5%).

Based on linear regression analyses (p<0.0001) among OC, EC, nss-K$^+$ and WSOC, the characteristic ratios of OC/EC, nss-K$^+$/OC and WSOC/OC are constrained for the emissions from agricultural-residue burning (Table 1, Fig. 3). The OC/EC ratio of 10.6 for paddy-residue burning emission is significantly higher than that for the wheat-residue burning emission (3.0; Fig. 3a). The nss-K$^+$/OC ratio of 0.06 for the paddy-residue burning emission is also different, lower by a factor of 2 than that from the wheat-residue burning emission (0.14; Fig. 3b). This is attributable to the high abundance of OC from paddy-residue burning emission. Furthermore, a major fraction of OC is found to be soluble in water from both the biomass burning emissions: the WSOC/OC ratio of 0.52 for paddy-residue is lower than that for the wheat-residue burning emission (0.60; Fig. 3c). In this context, an earlier study reports that the WSOC can have primary production from biomass burning emission as well as secondary formation pathway via photochemical reactions in the atmosphere (Mayol-Bracero et al., 2002). It is important to state here that a recent study from the Amazonian forest has suggested the significance of biogenic potassium to serving as a seed for SOA formation (Pöhlker et al., 2012). Thus, the SOA production can be favoured by their condensation onto K^+ particles following the photochemical reactions of volatile organic compound (VOC) with atmospheric oxidants. In our context, it is likely possible that the biomass burning derived potassium could also serve as a seed for the SOA production. Relatively high abundances of nss-K^+ and OC in $PM_{2.5}$ from paddy-residue burning emission would facilitate higher contribution of secondary organic carbon (SOC) than that from the wheat-residue burning (Fig. 3b; Table 1). Furthermore, during wintertime (December–March) the OC with EC and OC with nss-K^+ suggests for the dominant impact from biomass burning emission in the IGP [OC/EC: 4.2; nss-K^+/OC: 0.0.06]. The WSOC/OC ratio (0.62) in winter looks also similar to that observed from two distinct post-harvest biomass burning emissions (of paddy- and wheat-residue).

PAH isomer ratios for the fossil-fuel combustion, forest fires and bio-fuels burning emission are available in the literature (Kirton et al., 1991; Khalili et al., 1995; Masclet et al., 1995; Schauer et al., 2001; Sheesley et al., 2003; Khillare et al., 2005a, 2005b). However, the information on PAH isomer ratios from the two potential sources of carbonaceous aerosols in the IGP, paddy- and wheat-residue burning (under ambient atmospheric conditions) is rather lacking in the literature (Rajput et al., 2011b). In this context, the present study serves to provide several PAH isomer ratios along with the major composition of carbonaceous species (EC, OC and WSOC) and nss-K^+ in $PM_{2.5}$ from paddy- and wheat-residue burning emission for statistically significant number of data set (Table 1). Employing the PAH isomers, ANTH/(ANTH+PHEN) and IcdP/(IcdP+BghiP) against FLA/(FLA+PYR) in the cross plots, we document a characteristic information for the post-harvest agricultural-waste burning emissions from paddy- and wheat-residues (Fig. 4a and b; Table 1), distinctly different from the fossil-fuel combustion sources. The chemical characteristics of carbonaceous aerosols from biomass burning and fossil-fuel combustion sources in the IGP (from Patiala) during wintertime are summarised in Table 1 (for n=51).

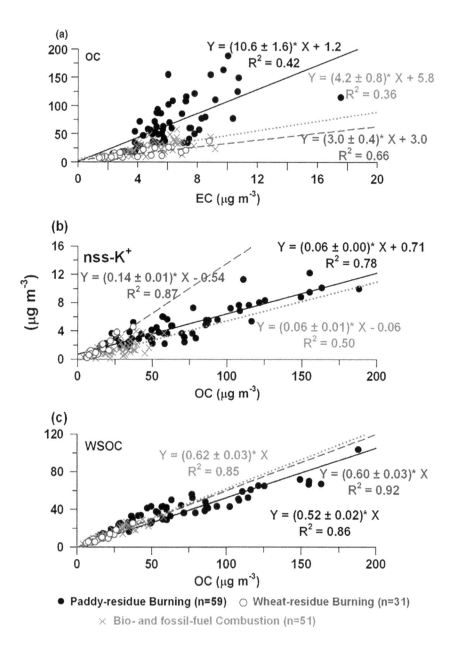

FIGURE 3: Scatter plots of (a) EC vs. OC; (b) OC vs. non-sea-salt: nss-K+; and (c) OC vs. WSOC during different emissions in the IGP.

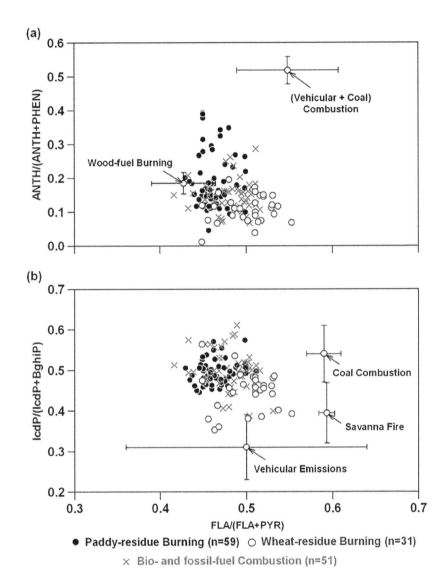

FIGURE 4: Cross plots of PAH isomers showing distinct differences for the biomass burning emission vis-à-vis fossil-fuel combustion in the IGP: (a) FLA/(FLA+PYR; 4-ring on X-axis) vs. ANTH/(ANTH+PHEN; 3-ring on Y-axis); and (b) FLA/(FLA+PYR; 4-ring on X-axis) vs. IcdP/(IcdP+BghiP; 6-ring on Y-axis). Other data source: wood-fuel burning (Bari et al., 2009); (Vehicular+Coal) combustion (Sharma et al., 2008); vehicular emission (Khillare et al., 2005a, b; Rajput and Lakhani, 2008); coal combustion (Kirton et al., 1991; Khalili et al., 1995; Li et al., 2010); savanna fire (Masclet et al., 1995).

TABLE 2: Spatial distribution of OC/EC ratio in ambient aerosols from the Indo-Gangetic Plain (IGP) and different locations during wintertime (December–March)

Location	N	Concentration ($\mu g\ m^{-3}$)	OC (%)	EC (%)	OC/EC	
IGP						
Patiala	51	124±58 ($PM_{2.5}$)	22±4	4.6±1.8	4.2±0.8	This study
Hisar	42	177±64 (TSP)	17	2.1	8.5±2.2	Rengarajan et al., 2007
Kanpur	17	141±73 (TSP)	18	3.4	6.2±3.7	Ram and Sarin, 2010
Allahabad	19	300±90 (TSP)	16	2.1	8.1±1.7	
Western India						
Ahmedabad	16	85±37 ($PM_{2.5}$)	33±6	6±1	6.1±1.2	This study
Mt Abu	15	22±9 ($PM_{2.5}$)	10±6	3.3±1.6	3.0±1	This study
Mumbai	4	128±27 (PM_{10})	20	10	2.0±0.3	Venkataraman et al., 2002
Southern India						
Chennai	29	(PM_{10})[a]			1.5±0.5	Pavuluri et al., 2011
Himalaya						
NE-Himalaya	51	97±50 ($PM_{2.5}$)	36±8	6±3	6.8±3.2	Rajput et al., 2013
NCO-P	28	(PM_{10})[a]			11±2	Decesari et al., 2010
Manora peak	38	66±18 (TSP)	16	3	6.3±2.2	Ram et al., 2010
Ocean						
Northern-BoB	31	38±20 ($PM_{2.5}$)	16	5	3	Srinivas et al., 2011
Southern-BoB		22.3±9.9 ($PM_{2.5}$)	12	5	2.2	

[a]PM10 concentration is not reported (NR).
TSP refers to total suspended particulate matter.

12.3.2 OC/EC RATIO DURING WINTERTIME (DECEMBER–MARCH) FROM DIFFERENT GEOGRAPHICAL LOCATIONS IN INDIA

Earlier studies conducted in the IGP have reported on the concentrations of OC and EC in total suspended particulate matter (TSP; Table 2) during the wintertime (December–March) (Rengarajan et al., 2007; Ram and Sarin, 2010). The spatial variability in average concentration of TSP was reported to be ~100–400 during the wintertime in the IGP. The average mass fractions of OC and EC are nearly identical varying from ~16–18 and 2–4%, respectively, at study locations (Table 2). Based on the aerosol composition analysis from December 2008 to March 2009 and December 2010 to March 2011 at Patiala in the IGP (this study), we find $PM_{2.5}$ concentrations varying from 19 to 244 (124±58) µg m^{-3} (Tables 1 and 2), quite similar to TSP from other locations in the IGP. The mass fractions of OC (22±4%) and EC (4.6±1.8%) in $PM_{2.5}$ at Patiala are also quite similar to that in TSP over other locations (Hisar, Kanpur and Allahabad) during the wintertime. Thus, contribution of fine-mode aerosols (aerodynamic diameter $\leq_{2.5}$ µm) to TSP is dominant during the wintertime. The major source of particulate matter during the wintertime in the IGP is attributed to emissions from bio-fuel burning and fossil-fuel combustion sources (Rengarajan et al., 2007; Ram and Sarin, 2010; Rajput et al., 2011b).

Furthermore, downwind transport of pollutants from IGP along the foot-hills of NE-Himalaya, during the wintertime, has been documented by a recent study from NE-Himalaya [Barapani: 25.7°N, 91.9°E; 1064 m amsl; n=51, January–March 2009 and 2010], wherein $PM_{2.5}$ concentration varies from 39 to 348 (Av: 97±50) µg m^{-3}: of which OC and EC contributes 36±8 and 6±3%, respectively (Rajput et al., 2013). The OC/EC ratio ~7.0 from the foot-hills of NE-Himalaya is similar to that observed from several locations in the IGP. In contrast, over northern Bay of Bengal (N-BoB: ~10–20°N latitude; n=31) the $PM_{2.5}$ concentration in the marine atmospheric boundary layer (MABL) varies from 13.2 to 76.7 (38±20) µg m^{-3}; with contributions from OC and EC as 16 and 5%, respectively (Srinivas et al., 2011). The $PM_{2.5}$ concentration over the southern BoB (S-BoB: ~2–10°N latitude) varies from 2.0 to 35.3 (22.3±9.9) µg m^{-3}; with contributions from OC and EC as 12 and 5%, respectively. Thus, OC mass fraction decreases, whereas EC contribution remains constant as a function of distance from the source region (with north-south gradient over open ocean). This is attributable to the preferential removal of OC (residing in coarse fraction) compared to EC (in fine fraction) in the MABL over the BoB. A systematic decrease in OC/EC ratio has been also observed over the Pacific in the outflow from East Asia (Lim et al., 2003). They attributed the decrease in OC/EC ratio to relatively longer atmospheric life-time of EC than OC in the MABL.

In order to assess the spatial distribution of carbonaceous aerosol in terms of OC/EC ratio from different geographical locations over India, aerosol sampling during wintertime from two additional sites, at Ahmedabad [23.03°N, 72.65°E; 49 m amsl; n=16, 1 December 2009–28 January 2010] and Mt Abu [24.6°N, 72.7°E; 1680 m amsl; n=15, 12 February–17 March 2010] in semi-arid region of western India, has been conducted in this study. The PM$_{2.5}$ concentration varied from 32 to 161 (Av±sd: 85±37) µg m^{-3}, of which 33±6% is OC and 6±1% is EC at Ahmedabad. The PM$_{2.5}$ concentration varied from 10 to 38 (22±9) µg m^{-3}, of which 10±6% is OC and 3.3±1.6% is EC at Mt Abu. It is relevant to state that during wintertime, prevailing NE-winds favour the long-range transport of pollutants from the IGP to western India. Thus, aerosols over Ahmedabad represent regional characteristics as well as components from long-range transport. However, the other site in western India at Mt Abu (1680 m amsl) is by-and-large impacted by the long-range transport.

The OC/EC average ratio as high as ~6–10 in the upwind locations in the IGP (Table 2) is attributable to the dominance of biomass burning emission sources and photochemical reactions (Rengarajan et al., 2007; Ram and Sarin, 2010; Rajput et al., 2011a). A similar OC/EC ratio of ~6 from Manora Peak and ~10 in aerosols from the southern slope of higher Himalaya (National Climate Observatory-Pyramid: NCO-P sampling station) have been also attributed earlier to the dominance of biomass burning emissions and/or contributions from SOA (Decesari et al., 2010; Ram et al., 2010). Furthermore, a similar OC/EC ratio from a semi-arid location at Ahmedabad (~6) is observed during wintertime. In contrast, at other places in India (Mumbai and Chennai), the OC/EC ratio is ~2–3 (Venkataraman et al., 2002; Pavuluri et al., 2011). However, the OC/EC ratio in winter over the BoB has been reported to be >2 (Srinivas et al., 2011). Thus, we integrate the present-day wintertime (December–March) scenario on significant variability in the chemical characteristics of carbonaceous aerosols in terms of the OC/EC ratio from the IGP and those over different environmental regions in and around the country.

TABLE 3: Emission budget of carbonaceous species from post-harvest agricultural-waste burning in the Indo-Gangetic Plain (IGP)

Open biomass Paddy-residue burning (October–November)	Emission factor[a] (g/kg)	Fuel load[b] (Kg/sq. km)	Area burnt[c] (sq. km/y)	Emission budget[d]
OC	7.6±1.2	11.8×10⁵	48 400	436±68 Gg/y
EC	0.72±0.03			41±2 Gg/y
ΣPAHs	2.8±0.5[e]			161±31 Mg/y
Wheat-residue burning (April–May)				
OC	1.2±0.03	5.94×10⁵	48 400	69±2 Gg/y
EC	0.31±0.02			18±1 Gg/y
ΣPAHs	0.4±0.1e			21±6 Mg/y

[a]*EF modified, after (Kanokkanjana et al., 2011) for paddy-residue burning and, after (Hays et al., 2005) for wheat-residue burning.*
[b]*Adopted from a study in the IGP (Badarinath et al., 2006).*
[c]*Inferred from MODIS (Aqua/Terra) satellite data (Resolution: 1°×1° lat.–long. grid).*
[d]*Emission factor×fuel load×area.*
[e]*Emission factor in mg/kg.*

12.3.3 EMISSION BUDGET OF CARBONACEOUS SPECIES FROM PADDY- AND WHEAT-RESIDUE BURNING IN THE IGP

The emission of carbonaceous species (EC, OC and ΣPAHs) from the paddy- and wheat-residue burning in the IGP is estimated using the following equation:

$$\text{Emission (Gg/y)} = \text{EF (g/kg)} * \text{FL (kg/km}^2) * \text{AB (km}^2.\text{y}) * 10^{-9} \quad (1)$$

Here, 'EF' is abbreviated for emission factor; 'FL' for fuel load over the agricultural fields and 'AB' for annual area burnt in the IGP (Table 3). A factor of 10^{-9} is multiplied to convert gram-emissions into Giga-grams (Gg).

In this study, the EF for carbonaceous species (EC, OC and ΣPAHs) from post-harvest paddy- and wheat-residue burning is suitably adopted from recent studies (Hays et al., 2005; Kanokkanjana et al., 2011). We have taken into consideration the differences in the OC/$PM_{2.5}$ and EC/ $PM_{2.5}$ ratios for the biomass burning under ambient atmospheric conditions and those based on chamber experiments in order to assess the emission factors (Hays et al., 2005; Kanokkanjana et al., 2011). In case of paddy-residue burning emissions, the average mass fraction of (OC+EC)/ $PM_{2.5}$ is 36.5$\pm$7.1% (this study). This mass fraction is in close agreement with that reported (38$\pm$2%) in a recent study (Kanokkanjana et al., 2011) for moist combustion of paddy-residue from irrigated fields. Therefore, we have adopted the EF of EC as 0.72$\pm$0.03 g/kg reported by (Kanokkanjana et al., 2011). The EF of OC is not assessed in their study on paddy-residue burning emissions, unlike our approach using the following equation:

$$EF_{OC} \text{ (g/kg)} = OC/EC * EF_{EC} \text{ (g/kg)} \tag{2}$$

And, the EF of ΣPAHs for paddy-residue burning is estimated from the mass fraction of PAH (ΣPAHs/OC), using the following equation:

$$EF_{PAHs} \text{ (mg/kg)} = (\Sigma PAHs \text{ (mg))}/OC \text{ (g)} * EF_{OC} \text{ (g/kg)} \tag{3}$$

For wheat-residue burning emissions, the mass fraction of OC (26$\pm$5%) in the IGP is quite similar to that reported from a chamber experiment (Hays et al., 2005). However, the contribution of EC (6.9$\pm$2.5%) in $PM_{2.5}$ from wheat-residue burning emission in the IGP is about 1.6 times lower as compared to 11% in the chamber (Hays et al., 2005). Therefore, we have used the same EF for OC and scaled down the EF of EC by a factor of 1.6 as reported by (Hays et al., 2005) for wheat-residue burning emission under ambient atmospheric conditions (Table 3). The EF of ΣPAHs for paddy-residue burning is estimated from eq. (3).

The value of fuel load (Table 3) for the paddy- and wheat-residues over the agricultural fields is adopted from a recent literature representing the agricultural-scenario for the states of Punjab in the IGP (Badarinath et al., 2006).

Based on the analysis of open fire-counts satellite data (Resolution: 1°×1°, latitude-longitude grid) from the MODIS (on-board Aqua/Terra; level 2) during 2008–2011, the fire active (agricultural-waste burning) area in the IGP (Fig. 1; including the states of Punjab, Haryana and western part of UP) is estimated to be 48400 sq. km (Justice et al., 2002). The fire-count data (Fig. 1) has been corrected for Cloud and Overpass. The 'Cloud and Overpass corrected fire-pixel count' represents the number of pixels corrected for multiple satellite overpass, missing observations and variable cloud cover (Justice et al., 2002). Since the region (upwind IGP) is fire active for a total of 4 months in a year: due to the practice of paddy-residue burning during October–November and wheat-residue burning during April–May. Therefore, it is considered logical to estimate the emission of aerosols in the IGP from the paddy- and wheat-residue burning only for 2 months each, and represented here as the total emission (Table 3).

12.3.4 A GLOBAL SCENARIO ON BIOMASS BURNING EMISSIONS

Recently the global emission budget of EC and OC has been revised (Bond et al., 2013). Accordingly, ~62 Tg/y of OC (primary) and ~14 Tg/y of EC are emitted from the biomass burning and fossil-fuel combustion sources over the globe. Between the two sources, emission from biomass burning is dominant: producing ~89% of the total OC and 60% of the total EC. Furthermore, among the biomass burning sources, the major source of OC (26 Tg/y) and EC (4.3 Tg/y) is the bio-fuel burning (Fig. 5). The emission from savanna fires produces ~17 Tg/y of OC and 2.2 Tg/y of EC. The forest fires produce ~17 Tg/y of OC and 1.5 Tg/y of EC. The agricultural-waste burning emissions are reported to produce 1154 Gg/y of OC and 280 Gg/y of EC on a global scale. In this context, the net emissions of OC and EC from agricultural-waste burning from the IGP (Northern India) are estimated as 505±68 and 59±2 Gg/y, respectively (Table 3). Using the EC-tracer method (Castro et al., 1999), it has been estimated that ~50% of the OC is primary during both the paddy- and wheat-residue burning emissions. Thus, the net emission of primary OC from agricultural-waste (paddy- and wheat-residue) burning is estimated as 252±34 Gg/y. As far as the emission budgets from the agricultural-waste burning on a global scale are concerned, ~22% of primary OC [252±34 Gg/y] and 21% of EC [59±2 Gg/y] are produced from the IGP (Northern India; Fig. 5). However, on a global biomass burning emission scale, as of present understanding, the contribution of primary OC and EC from the agricultural-waste burning emission is 2 and 3%, respectively (Bond et al., 2013).

12.4 CONCLUSIONS AND IMPLICATIONS

Large-scale emissions from paddy-residue burning during October–November and wheat-residue burning in April–May are conspicuous features in the IGP. We document significant differences in the OC/EC, nss-K+/OC, WSOC/OC, ΣPAHs/EC and PAH isomer ratios for the two biomass burning sources. Relatively high emissions of OC, EC and PAHs are associated with the paddy-residue burning compared to that from wheat-residue burning emissions. The high abundance of OC, shallow boundary layer height during the wintertime and secondary formation of organic aerosols contribute to the fog and haze conditions over Northern India. This also addresses the issue of over projecting the role of black carbon in the atmospheric radiative forcing over Northern India. The large seasonal variability in aerosol composition associated with varying biomass burning emissions vis-à-vis fossil-fuel combustion sources in the IGP have implications to heterogeneous-phase chemistry of organic aerosols and oxidants (O_3 and OH radical).

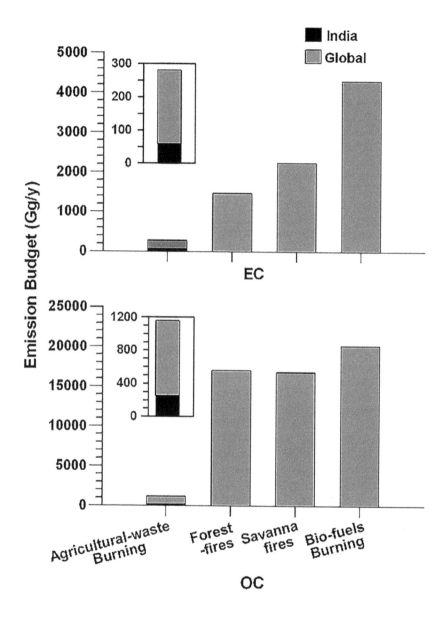

FIGURE 5: Emission budgets of aerosol OC (primary) and EC from different biomass burning over the globe. Our data put together with a recent emission inventory suggest that OC and EC produced from agricultural-waste burning in the IGP (Northern India; shown as inset) contributes to ~22% [primary OC: 252±34 Gg/y] and 21% [EC: 59±2 Gg/y], respectively, on the global agricultural-waste burning emission scale.

REFERENCES

1. Badarinath, K. V. S., Chand, T. R. K. and Prasad, V. K. 2006. Agricultural crop residue burning in the Indo-Gangetic Plains – a study using IRS-P6 A WiFS satellite data. Curr. Sci. 91, 1085–1089.
2. Bari, M. A., Baumbach, G., Kuch, B. and Scheffknecht, G. 2009. Wood smoke as a source of particle-phase organic compounds in residential areas. Atmos. Environ. 43, 4722–4732.
3. Birch, M. E. and Cary, R. A. 1996. Elemental carbon-based method for monitoring occupational exposures to particulate diesel exhaust. Aerosol Sci. Technol. 25, 221–241.
4. Bond, T. C., Doherty, S. J., Fahey, D. W., Forster, P. M., Berntsen, T. and co-authors. 2013. Bounding the role of black carbon in the climate system: a scientific assessment. J. Geophys. Res. DOI: 10.1002/jgrd.50171.
5. Castro, L. M., Pio, C. A., Harrison, R. M. and Smith, D. J. T. 1999. Carbonaceous aerosol in urban and rural European atmospheres: estimation of secondary organic carbon concentrations. Atmos. Environ. 33, 2771–2781.
6. Decesari, S., Facchini, M. C., Carbone, C., Giulianelli, L., Rinaldi, M. and co-authors. 2010. Chemical composition of PM10 and PM1 at the high-altitude Himalayan station Nepal Climate Observatory-Pyramid (NCO-P) (5079 m a.s.l.). Atmos. Chem. Phys. 10, 4583–4596.
7. Gupta, P. K., Sahai, S., Singh, N., Dixit, C. K., Singh, D. P. and co-authors. 2004. Residue burning in rice-wheat cropping system: causes and implications. Curr. Sci. 87, 1713–1717.
8. Hays, M. D., Fine, P. M., Geron, C. D., Kleeman, M. J. and Gullett, B. K. 2005. Open burning of agricultural biomass: physical and chemical properties of particle-phase emissions. Atmos. Environ. 39, 6747–6764.
9. Justice, C. O., Giglio, L., Korontzi, S., Owens, J., Morisette, J. T. and co-authors. A. 2002. The MODIS fire products. Remote Sens. Environ. 83, 244–262.
10. Kanokkanjana, K., Cheewaphongphan, P. and Garivait, S. 2011. Black carbon emission from paddy field open burning in Thailand. IPCBEE Proc. 6, 88–92.
11. Keene, W. C., Pszenny, A. A. P., Galloway, J. N. and Hawley, M. E. 1986. Sea-salt corrections and interpretation of constituent ratios in marine precipitation. J. Geophys. Res. 91, 6647–6658.
12. Khalili, N. R., Scheff, P. A. and Holsen, T. M. 1995. PAH source fingerprints for coke ovens, diesel and, gasoline engines, highway tunnels, and wood combustion emissions. Atmos. Environ. 29, 533–542.
13. Khillare, P., Balachandran, S. and Hoque, R. 2005a. Profile of PAH in the exhaust of gasoline driven vehicles in Delhi. Environ. Monit. Assess. 110, 217–225.
14. Khillare, P. S., Balachandran, S. and Hoque, R. R. 2005b. Profile of PAHs in the diesel vehicle exhaust in Delhi. Environ. Monit. Assess. 105, 411–417.
15. Kirton, P. J., Ellis, J. and Crisp, P. T. 1991. The analysis of organic matter in coke oven emissions. Fuel. 70, 1383–1389.

16. Li, P.-h., Wang, Y., Li, Y.-H., Wang, Z.-F., Zhang, H.-Y. and co-authors. 2010. Characterization of polycyclic aromatic hydrocarbons deposition in $PM_{2.5}$ and cloud/fog water at Mount Taishan (China). Atmos. Environ. 44, 1996–2003.

17. Lim, H. J., Turpin, B. J., Russell, L. M. and Bates, T. S. 2003. Organic and elemental carbon measurements during ACE-Asia suggest a longer atmospheric lifetime for elemental carbon. Environ. Sci. Technol. 37, 3055–3061.

18. Masclet, P., Cachier, H., Liousse, C. and Wortham, H. 1995. Emissions of polycyclic aromatic hydrocarbons by savanna fires. J. Atmos. Chem. 22, 41–54.

19. Mayol-Bracero, O. L., Guyon, P., Graham, B., Roberts, G., Andreae, M. O. and co-authors. 2002. Water-soluble organic compounds in biomass burning aerosols over Amazonia 2. Apportionment of the chemical composition and importance of the polyacidic fraction. J. Geophys. Res. 107, 8091.

20. Pavuluri, C. M., Kawamura, K., Aggarwal, S. G. and Swaminathan, T. 2011. Characteristics, seasonality and sources of carbonaceous and ionic components in the tropical aerosols from Indian region. Atmos. Chem. Phys. 11, 8215–8230.

21. Pöhlker, C., Wiedemann, K. T., Sinha, B., Shiraiwa, M., Gunthe, S. S. and co-authors. 2012. Biogenic potassium salt particles as seeds for secondary organic aerosol in the Amazon. Science. 337, 1075–1078.

22. Punia, M., Nautiyal, V. P. and Kant, Y. 2008. Identifying biomass burned patches of agricultural residue using satellite remote sensing data. Curr. Sci. 94, 1185–1190.

23. Rajput, N. and Lakhani, A. 2008. Measurements of polycyclic aromatic hydrocarbons at an industrial site in India. Environ. Monit. Assess. 150, 273–284.

24. Rajput, P., Sarin, M. M. and Kundu, S. S. 2013. Atmospheric particulate matter ($PM_{2.5}$), EC, OC, WSOC and PAHs from NE-Himalaya: abundances and chemical characteristics. Atmos. Poll. Res. 4, 214–221.

25. Rajput, P., Sarin, M. M. and Rengarajan, R. 2011a. High-precision GC-MS analysis of atmospheric polycyclic aromatic hydrocarbons (PAHs) and isomer ratios from biomass burning emissions. J. Environ. Prot. 2, 445–453.

26. Rajput, P., Sarin, M. M., Rengarajan, R. and Singh, D. 2011b. Atmospheric polycyclic aromatic hydrocarbons (PAHs) from post-harvest biomass burning emissions in the Indo-Gangetic Plain: isomer ratios and temporal trends. Atmos. Environ. 45, 6732–6740.

27. Ram, K. and Sarin, M. M. 2010. Spatio-temporal variability in atmospheric abundances of EC, OC and WSOC over Northern India. J. Aerosol Sci. 41, 88–98.

28. Ram, K. and Sarin, M. M. 2011. Day-night variability of EC, OC, WSOC and inorganic ions in urban environment of Indo-Gangetic Plain: implications to secondary aerosol formation. Atmos. Environ. 45, 460–468.

29. Ram, K., Sarin, M. M. and Hegde, P. 2010. Long-term record of aerosol optical properties and chemical composition from a high-altitude site (Manora Peak) in Central Himalaya. Atmos. Chem. Phys. 10, 11791–11803.

30. Ram, K., Sarin, M. M. and Tripathi, S. N. 2012. Temporal trends in atmospheric $PM_{2.5}$, PM10, elemental carbon, organic carbon, water-soluble organic carbon, and optical properties: impact of biomass burning emissions in the Indo-Gangetic Plain. Environ. Sci. Technol. 46, 686–695.

31. Ramanathan, V., Li, F., Ramana, M. V., Praveen, P. S., Kim, D. and co-authors. 2007. Atmospheric brown clouds: hemispherical and regional variations in long-range transport, absorption, and radiative forcing. J. Geophys. Res. 112, D22S21.

32. Rengarajan, R., Sarin, M. M. and Sudheer, A. K. 2007. Carbonaceous and inorganic species in atmospheric aerosols during wintertime over urban and high-altitude sites in North India. J. Geophys. Res. 112, D21307.

33. Schauer, J. J., Kleeman, M. J., Cass, G. R. and Simoneit, B. R. T. 2001. Measurement of emissions from air pollution sources. 3. C1 – C29 organic compounds from fireplace combustion of wood. Environ. Sci. Technol. 35, 1716–1728.

34. Sharma, H., Jain, V. and Khan, Z. 2008. Atmospheric polycyclic aromatic hydrocarbons (PAHs) in the urban air of Delhi during 2003. Environ. Monit. Assess. 147, 43–55.

35. Sheesley, R. J., Schauer, J. J., Chowdhury, Z., Cass, G. R. and Simoneit, B. R. T. 2003. Characterization of organic aerosols emitted from the combustion of biomass indigenous to South Asia. J. Geophys. Res. 108, 4285.

36. Srinivas, B., Kumar, A., Sarin, M. M. and Sudheer, A. K. 2011. Impact of continental outflow on chemistry of atmospheric aerosols over tropical Bay of Bengal. Atmos. Chem. Phys. Discuss. 11, 20667–20711.

37. Venkataraman, C., Reddy, C. K., Josson, S. and Reddy, M. S. 2002. Aerosol size and chemical characteristics at Mumbai, India, during the INDOEX-IFP (1999). Atmos. Environ. 36, 1979–1991.

Author Notes

CHAPTER 2

Acknowledgments

The present work has been carried out within the activities of the Project Line no.3 of the MONITER Project. The project was funded by the Health and Environmental Assessorships of the Emilia-Romagna Region. The authors thank Dr. Paola Angelini and all project participants.

CHAPTER 3

Funding

This research received no specific grant from any funding agency in the public, commercial, or not for- profit sectors.

CHAPTER 4

Disclaimer

This report is a work commissioned by the National Institute of Health Research. The views expressed in this publication are those of the authors and not necessarily those of the NHS, the National Institute for Health Research, or the Department of Health.

Conflict of Interest

All authors declare no conflict of interests with any trademarks mentioned in this paper.

Acknowledgments

The work of the Small Area Health Statistics Unit is funded by the Public Health England as part of the MRC-PHE Centre for Environment and Health, funded also by the UK Medical Research Council, and held jointly between Imperial College London and King's College London. The authors thank the funders of the national municipal solid waste incinerators study: Public Health England and SAHSU. Danielle Ashworth is funded by an MRC-PHE Centre for Environment and Health PhD studentship. They thank Nick Bettinson at the Air Quality Modelling Assessment Unit at the Environment Agency for advice on dispersion modelling and the use of ADMS. The authors thank the Environment Agency in England and Wales for providing data on incinerators and for their dispersion modelling advice. They thank the Office for National Statistics for providing census data and the Met Office/BADC for providing meteorological data. Paul Elliott acknowledges support from the National Institute for Health Research (NIHR) Biomedical Research Centre at Imperial College Healthcare NHS Trust and Imperial College London. Paul Elliott is an NIHR Senior Investigator.

CHAPTER 5

Author Contributions
All authors contributed extensively to the work presented in this paper.

Conflict of Interest
The authors declare no conflict of interest.

CHAPTER 6

Note
This paper was made thanks to the collaboration and financing of the Autonomous Province of Trento. The Authors acknowledge also to the local Environmental Protection Agency, and the Municipality where the plant is located, for the collaboration in the organization of the sampling campaign related to the present paper.

CHAPTER 8

Acknowledgments

This research was funded by the RAND Corporation's Investment in People and Ideas program. Support for this program is provided, in part, by the generosity of RAND's donors and by the fees earned on client-funded research. We thank Pennsylvania Department of Environmental Protection (PA DEP) staff and one shale gas operator company for providing data and helpful discussions. Joe Osborne of the Group Against Smog and Pollution (GASP) provided access to compiled compressor station permits, and Nick Muller (Middlebury College) assisted with understanding and making use of the APEEP model. Thanks also to Henry Willis, Shanthi Nataraj, and Tom LaTourrette of RAND for helpful discussions. We also thank two anonymous reviewers for suggestions that significantly improved the manuscript.

CHAPTER 9

Acknowledgments

Traffic data were provided by the Municipality of Verona, in particular by Mr B. Pezzuto, whose helpfulness is gratefully acknowledged. The Authors wish to thank the Regional Environmental Agency of Veneto (ARPA Veneto, Department of Verona) for the meteorological data. The authors are also thankful to Mr. A. Piovesan of Azienda Trasporti Verona Srl for the information about the composition of the municipal bus fleet. Special thanks to the Fondazione Trentina per la Ricerca sui Tumori and, especially, to the De Luca family for the financial support to the research activity

CHAPTER 10

Acknowledgements

This work was funded by the Spanish Ministry of Science and Innovation (VAMOS CGL2010-19464/CLI; DAURE CGL2007-30502-E/CLI,

GRACCIE- CSD2007-00067), Department of Inovation, Science and Enterprise of the Andalusian Autonomous Government (AER-REG-P07-RNM-03125), the Ministry of the Environment and Rural and Marine Affairs, and the 7th FP from the EC project SAPUSS (Marie Curie intra-European Fellowship). The authors acknowledge the Departament de Territori i Sostenibilitat from Generalitat de Catalunya, Gobierno de Canarias and Junta de Andaluc´ıa (Spain), DEFRA (UK) and the Swiss Federal Office for the Environment (FOEN) for providing the data.

CHAPTER 11

Acknowledgments

We thank Gary Norris, Carry Croghan and Rich Cook at US EPA, Laprisha Berry Vaughn, Sonya Grant, Chris Godwin, Graciela Mentz, Xiaodan Ren, Irme Cuadros and other staff at the University of Michigan, and Brian Naess, Mohammad Omary, Kevin Talgo, Alejandro Valencia, Yasuyuki Akita and Marc Serre of the University of North Carolina at Chapel Hill. We are grateful to the NEXUS participants and their families who assisted us with the collection of these data. Community Action Against Asthma is a community-based participatory research partnership aimed at investigating the influence of environmental factors on childhood asthma. We acknowledge the contributions of all of the partners involved in this collaborative effort: Arab Community Center for Economic and Social Services; Community Health & Social Services Center; Detroit Department of Health and Wellness Promotion; Detroit Hispanic Development Corporation; Detroiters Working for Environmental Justice; Friends of Parkside; Latino Family Services; Southwest Detroit Environmental Vision; Warren/Conner Development Coalition; Institute for Population Health, and the University of Michigan Schools of Public Health and Medicine. The US Environmental Protection Agency through its Office of Research and Development partially funded the research described here under cooperative agreement R834117 (University of Michigan). It has been subjected to Agency review and approved for publication.

Author Contributions

Vlad Isakov conceptualized the analysis, coordinated contributions from the team, produced drafts and coordinated revisions of the paper. Saravanan Arunachalam led the modeling efforts, including model setup, model simulations and evaluation. Michelle Snyder led the R-LINE model development, and performed evaluation. Janet Burke contributed to the analysis of exposure metrics, coordinated input from the NEXUS team, edited and helped to revise the paper. Stuart Batterman contributed to the development and evaluation of the modeled exposure metrics and edited the paper. Kathie Dionisio contributed to analyses of the exposure metrics and edited the paper, and Sarah Bereznicki contributed to analyses of the NEXUS measurements. David Heist, Steve Perry, Val Garcia and Alan Vette contributed to the design of the modeling study and analyses of the exposure metrics. All authors read and approved the final manuscript.

Conflict of Interest

The authors declare no conflict of interest.

CHAPTER 12

Acknowledgments

We acknowledge the funding support received from Indian Space Research Organization-Geosphere Biosphere Program Office (Bengaluru, India). We are thankful to Punjab Agricultural University (Dr. Varinderpal Singh) for providing the data on moisture content of post-harvest crop-residues from agricultural fields in Punjab state (in the Indo-Gangetic Plain). We thank two anonymous reviewers for providing their constructive comments and suggestions and Dr. Kaarle Hämeri for editorial handling of the manuscript.

Index

Milton Keynes UK
Ingram Content Group UK Ltd.
UKHW022046141024
449569UK00022B/827